MEMORY CRAFT

MEMORY CRAFT

IMPROVE YOUR MEMORY WITH THE
MOST POWERFUL METHODS IN HISTORY

LYNNE KELLY

PEGASUS BOOKS
NEW YORK LONDON

MEMORY CRAFT

Pegasus Books, Ltd.
148 West 37th Street, 13th Floor
New York, NY 10018

First Pegasus Books hardcover edition January 2019

ISBN: 978-1-64313-324-9

10 9 8 7 6 5 4 3 2 1

Printed in the United States of America
Distributed by W. W. Norton & Company

For Damian Kelly
and Rebecca, Rudolph, Abigail and Leah Heitbaum.

And for the thousands of students who have been in my
classrooms over my long career. From you I learned so much.

CONTENTS

CONTENTS

CONTENTS

LIST OF FIGURES

INTRODUCTION

I was blessed with an appallingly bad memory. I say blessed, because if it wasn't for this fact I would have never asked the question that changed my life: 'How the hell did they remember so much stuff?'

About ten years ago, I wanted to write a book about animal behaviour and wondered how much more I might observe when watching birds and mammals and insects and my beloved spiders if I did so having read indigenous stories about these animals. La Trobe University had given me a PhD scholarship to research and write the book. My publisher was interested in publishing it. Lovely! Then I derailed the whole project by asking myself that simple question when I realised just how much practical stuff was stored in the memory of indigenous elders.

The Navajo kept a field guide to over 700 insects in memory. Only ten were critical, because they annoyed stock or messed with their crops. One they ate, the cicada. All the rest were known because the Navajo, like all humans, are curious and value knowledge for knowledge's sake. In the Navajo's stories the insects often act as metaphors to reflect upon human origins and behaviour. And that's just insects. Add the information about all the other animals, a thousand or so plants, complex genealogies, geology and astronomy, land management, navigation, timekeeping and weather and seasons . . . the list goes on and on.

The important lesson for all of those wanting to memorise huge amounts of information is that the Navajo store this knowledge in their mythology. In stories. Vivid lively stories make information more memorable.

And then it gets even more astounding. Recent research on Australian Aboriginal stories about landscape changes shows that the knowledge can be reliably dated to at least 7000 or even 10,000 years ago, and probably even longer. Ten thousand years ago is a really long time before the Egyptian pyramids and Stonehenge. Aboriginal stories tell exactly what happened to the landscape after the last ice age. I was astonished that they could retain information so accurately over such an incredibly long passage of time.

Indigenous cultures relied on their memories to store all the information on which they depended, both physically and culturally. Right into the Middle Ages and Renaissance, Western students were also taught to train their memories. It is only in the last few hundred years of Western society that we have lost the ability to memorise vast amounts of information. We use writing and technology to do the job for us. But memory, writing and technology can all enhance each other. This book is about how to do just that. I want to convince you that learning the memory arts is hugely worthwhile and great fun.

The memory methods used by indigenous cultures the world over have a great deal in common. If these memory methods are so effective and universal, they must be directly related to the way the brain works. It is the only common factor. And even more exciting, especially for someone over 60, as I am, is that current neuroscience research on the plasticity of the brain indicates that it doesn't have to decline with age.

I am convinced that we are very much poorer for not using our memories effectively anymore. Could this be a contributing factor to the prevalence of dementia and the general acceptance that memory fades with age? Or does it in fact fade with lack of use?

Early in my university research as a mature age student, I happened to visit Stonehenge with my husband, Damian, who was studying archaeology. I knew very little about it except that non-literate cultures—who had built ancient monuments all round the world, including Stonehenge, Easter Island moai, the Nasca Lines and the many monumental sites across the Americas—had always used memory methods closely related to the way the brain works. It became clear to me that a significant proportion of the purpose of sacred places was to do with memorising and conveying critical knowledge.

Nearly a decade later, PhD in hand, my ideas on indigenous cultures and archaeology were published by Cambridge University Press as *Knowledge and Power in Prehistoric Societies: Orality, memory and the transmission of culture* (2015). The thorough examination by archaeologists, anthropologists and memory experts of both my thesis and the resulting book gave me the authority to take these ideas to the mainstream reader with the book *The Memory Code* (2016). Interested though readers were in the new ideas about archaeology and anthropology, by far the greatest interest was in the extraordinary memory techniques described and how these could be applied in everyday life.

I started experimenting with a huge range of memory methods over ten years ago, and in the few years since these last two books were published those experiments have dominated my life. I have memorised more than I would have conceived possible about pre-history, history, geography, birds, mammals, trees and the complex spider families. I've memorised the periodic table, the history of writing, the history of timepieces and musical instruments. I have engaged with the lessons to be learned from the lives of the 130 historic figures I have chosen to be my 'ancestors'. I've used the memory methods to learn vocabulary for French and Mandarin and those difficult Chinese characters. I've memorised the history of art and attended classes to create contemporary works that reflect ancient memory techniques. I don't know everything about all of these topics—far from it. But I've memorised structures on which I can build more and more intricate understanding throughout the rest of my life.

Oh, and I now memorise shuffled decks of cards and long lists of random numbers for fun and competition. It really is fun, although I would never have believed it possible if I hadn't tried it for myself.

I'm not suggesting that you memorise the things that I have chosen. They are examples of how the vast range of memory techniques used throughout time and around the world can be implemented in our lives today.

That is what this book is about.

What you'll learn

Common sense would tell you that the advent of writing would destroy the need for memory methods. But memory methods persisted, taught in ancient Greek and Roman schools, in the monasteries of the Middle Ages and well into the Renaissance. It's from this time of early literacy that we get the easiest starting point for us to learn the memory arts. We'll start with the critical role of the imagination. I'll show you how to use your playful and creative inventiveness to memorise information with very straightforward medieval techniques.

I'll then build up to the more complex techniques from indigenous cultures, where you can start to choose which method, or even combination of methods, suits your particular need. Throughout the book, I'll give you examples of how I have implemented each of the techniques, just to make my explanations clearer. (Not because I expect you to memorise a field guide to the birds, Chinese or the implication of Young's double-slit experiment.) I've tried to choose a wide variety of examples so you can see how you will be able to adapt the methods to almost any form of knowledge. I expect you will have all sorts of other topics that you long to command.

The table of memory methods in Appendix A gives you a smorgasbord of memory methods and what sort of information they best suit. It's then up to you to choose.

In the first chapter, I'll introduce my version of a visual alphabet, a technique used during the later Middle Ages. It is a very straightforward system, for a list of items, that you should be able to use within half an hour. I'll then show the way a similar imaginative technique, a medieval bestiary, can be used for any words when they are not in a list or any particular order.

I mostly use my medieval bestiary to remember people's names, but it can be used to memorise anything that you can spell.

In Chapter 2, we'll get to the most powerful memory method of all, that of memory palaces. It is also known as the method of loci, the art of memory, memory journeys, songlines and many other terms. Memory is hugely enhanced by associating information with physical locations such as your home, neighbourhood or any other familiar place. Anything that you can put in order can be memorised using a memory palace.

Why do these memory methods work so well for everyone? In Chapter 3, I'll explain how these methods correlate with the most recent discoveries in neuroscience, which show that associating memory with place is hardwired into our brains. This common factor is why cultures all over the world have developed similar methods: they are working with the same brain structure. The neuroscience explains how we benefit from repetition and music, and in particular the value of memory palaces.

I had been repeatedly asked how to use memory methods to learn languages, my weakest area at school. So I decided to find out for myself and began to study the language I had dismally failed to learn at school, French. And then I decided to take on the massive challenge of learning Chinese. These two languages require very different approaches because they work in significantly different ways. I had no idea when I started. To finish Chapter 3, I'll take a quick look at applying memory methods in any academic study.

One of the most important lessons I have learned from indigenous cultures is the value of strong characters in stories.

I cannot emphasise enough how useful this is. You can repeat a string of facts over and over endlessly in your head but they'll disappear as soon as you stop that repetition. If you deploy lively characters to act out that knowledge you'll need far fewer repetitions to retain the knowledge, and you'll retain it for longer. It is why all indigenous cultures tell stories and why we should too. I'll explain it all in Chapter 4.

Indigenous cultures around the world don't just use the vast landscape as a memory palace; they use a wonderfully integrated system of objects—portable memory devices—that are often simply referred to as 'art' and seen to have little practical purpose.

When I started my research over a decade ago, I couldn't find a list of all the different memory devices anywhere. It was only by researching a huge range of cultures that I could see the similarities. In effect, many objects interpreted simply as artworks are mnemonic landscapes in miniature. I've tried lots of them, adapting here and there, and I cannot believe how effective they are, nor how beautiful. I'll explain in detail exactly how to create your own versions of the most effective devices I have found in Chapter 5.

We are all surrounded by the written word. But writing didn't happen instantaneously, it evolved slowly, and for a long time both written and oral memory methods worked hand in hand. In Chapter 6, I'll look at the evolution of writing and the way art provided a powerful memory device alongside the written word. We have gained so much from writing, but we've also lost. The good news is that we can take full advantage of our memories without losing any of the advantages that the written word, and now technology, offers.

In Chapter 7, I look at the medieval manuscripts and the gorgeous mnemonic art of the Renaissance. Once I stopped simply drooling over the extraordinary artistry of the medieval scribes, I was able to appreciate just how much their designs enhanced memory of the written word when books were so rare and a great deal of the population was still illiterate. The only thing duller than pages of neat and tidy handwritten notes is neat and tidy *typed* notes. If you want to remember what you've written down then take the lessons offered in the medieval manuscripts and turn your page into a memory space. Chapter 7 has a whole list of tricks for how to do this.

Readers of *The Memory Code* ask constantly about the implications of these memory techniques for education. I spent 40 years in classrooms—mostly in secondary schools, but also in primary schools and universities—and I only wish I'd known a great deal more about memory methods back then. Since writing *The Memory Code*, I have returned to the classroom, working with some superbly imaginative educators to explore the way memory methods can enhance what we already teach. They can be integrated beautifully into every area of the curriculum; you do not need to set up a separate subject. In every single classroom, we are using that very same human brain that neuroscience tells us is so perfectly adapted to memory techniques. So let's take advantage of it. That's the topic of Chapter 8.

At almost every talk I give on memory, someone will ask about the implications for old age and dementia. The neuroscience is clear: use it or lose it. Our entire identity is tied to our memories, and in losing them we cease to be ourselves. There's a great deal of promising research being done about using memory training with early-onset dementia patients. Valuable as that may be, every indication is that training our memories and setting ourselves up for old age could be invaluable for delaying, maybe even preventing, some forms of dementia. I'll explore the current state of the research and add a few of my own ideas in Chapter 9.

In *The Memory Code*, I said that I would never enter a memory championship because I couldn't handle the pressure. I was wrong. Not only did the memory training teach me how to handle pressure, it taught me how to concentrate and maintain focus. I now memorise shuffled decks of cards and strings of random numbers, despite the information being temporary and apparently useless. At the time of writing, I am Australia's Senior Memory Champion. Memory athletes populate an extraordinary world and, much to my surprise, I love being part of it. It is a rapidly growing sport for children, adults and even older people. I'll tell you all about that in Chapter 10.

Why memorise?

'Why memorise anything? We've got the internet now. We can just google it.'

I hear this so often that it scares me. I never hear: 'Why exercise? We can just order everything online. We don't have to use our muscles at all.' Our brain is a muscle. It needs exercise, even if we ignore every other benefit of using our memories.

I agree that there's a great deal we would have memorised in the past that need not be remembered today. I would never memorise a phone number or address anymore. I would not memorise information that I will only use once and can simply look up online.

But you can only look up information that you already know exists. When you look it up, you are burrowing down to specific factual knowledge. What I hadn't understood until I started committing data to memory, is that it is only with facts at my fingertips that I could play with information and see patterns I had never even glimpsed before.

It's only with a factual base that I believe we can use our higher levels of thinking effectively. If you're going to be creative, then you need to create with robust materials. Memory lays down the foundation on which you can build ever more complex layers of information and thinking and analysis.

But how do we distinguish what is worth memorising and what isn't? That's very much an individual choice. By the end of this book, I sincerely hope you will have decided exactly what you want on call from your memory and the best way to store it there.

CHAPTER 1

A medieval starting place

It is not advancing years that led to my forgetfulness. I have always been vague. Throughout my life I have put bits of paper and pens everywhere, ready to write down anything I needed to remember because I couldn't rely on my memory. I don't do that anymore; I trust my memory when I've encoded the information in a system. My natural memory is only marginally better than it ever was, but my trained memory can achieve so much more, even though I am at times horrified by the images my brain conjures up.

It is fortunate that no one knows what's going on in your imagination. The wilder, the more colourful and active, the more grotesque, vulgar or erotic the images and stories you create are, the more memorable they will be. That is the secret to making knowledge memorable.

The best place to start learning how to remember is a simple system that will let you practise creating vivid images. Don't worry if you think you lack a good imagination. I have yet to meet anyone who cannot rapidly develop the ability to create memorable pictures in their mind. You don't need to be a talented artist, brilliant writer or storyteller—you just need to imagine lots of action.

To memorise any information, you need to first organise it into little chunks that flow in a logical order. You then let your imagination make the link between the information you are trying to remember and some kind of vivid image that you can't forget. It may be slow at first but you will soon speed up.

We'll start with two memory methods from medieval Europe.[1] These were used alongside writing, so beautifully designed for those of us immersed in a written world. That makes them a simpler starting place than the more complex methods used by non-literate cultures.

Medieval memory arts

Early in the Middle Ages, the only books available were rare and precious handwritten manuscripts. There was no paper yet, so books were written on parchment made laboriously from animal skins, or on fragile papyrus. Both were very expensive materials. You could have used a wax tablet, but it needed to be constantly erased to use again. How could you possibly remember all that you were expected to know?

As in ancient Greek and Roman times, memory training was an essential part of the medieval curriculum. Medieval people were in awe of those with highly trained memories—it was not only a sign of intellect, as it had been in classical times, but also a mark of superior moral character. This is why many descriptions of the saints' lives mentioned their prodigious memory.

Slowly, as literacy took hold and books became available, highly illuminated pages of manuscripts and stylised artworks became memory devices for religious knowledge across what had been the Roman empire. Writing was considered an aid to memory, but most certainly not a substitute for it.

I'll explore the way the design of medieval manuscripts enhances memory in Chapter 7. At this stage, I just want to introduce two examples from that time to give you simple ways to use your imagination to make information memorable: visual alphabets and the bestiary.

Visual alphabets

A visual alphabet is a version of the peg system, in which you make some kind of link between what you want to remember and the object. You may be familiar with the most common peg system: associating an object with each of the numbers up to ten. Here's a common example:

1–sun	6–sticks
2–shoe	7–heaven
3–tree	8–gate
4–door	9–sign
5–hive	10–hen

If you want to remember to buy soap as your first item, then you imagine something linked to the sun, such as bright yellow soap that burns your fingers.

Instead of numbers, a visual alphabet uses letters. My alphabet starts with A–Arachne, B–bird of paradise, C–cat, D–dragon. It sounds harder than the rhyming 1–sun, 2–shoe . . . but I assure you that once you have the images in place it is easy to remember and more effective than using the numbers. More on that later.

Using the peg system for numbers given above, we hit a problem after 10–hen. The only thing I can find that rhymes with eleven is heaven and we've already used that for seven.

There are two further problems with this peg system that are immediately fixed by using a visual alphabet. Firstly, ten locations are too few. The alphabet gives us 26 locations. I have no trouble returning to A and using the letters again when I need to recall more than 26 items, because there is enough distance between the first A and the twenty-seventh A to avoid confusion. I rarely need a second pass.

Secondly, the suns, shoes and trees don't do much. They lack the character to excite your imagination. Instead of inanimate objects, I made my visual alphabet alive with dynamic creatures who interact with each other and with the objects to be remembered. They do more, so your imagination has more to play with.

I use my visual alphabet constantly. I use it for temporary information such as shopping lists and things I must remember to do. I use it when out birding. And I use it for public speaking, instead of any notes, which I'll explain below.

You can use a visual alphabet for any information that comes in a sequence; it doesn't have to be temporary. In the Middle Ages, they were used for permanent memorisation. Once your visual alphabet is established, which doesn't take long, you can use it forever. You will get better and faster at making the associations. You will find that recall from your memory is much faster than constantly finding your place on a written list.

Before we get to my visual alphabet and how to use it, it's worth having a look at the background to the design of my visual alphabet, which you can see in Plates 1–6.

Visual alphabets in the shapes of animals and humans were common in fifteenth-century writings about memory but are believed to date from much earlier. Medieval memory experts thought that the letters of the alphabet were too abstract, so they used animals or physical objects instead. The alphabet lends the familiar order so nothing is left out.

Some medieval visual alphabets were gloriously artistic, such as that of Milanese artist Giovannino de' Grassi. The letters of his alphabet, which he created around 1390, are shaped from the contorted figures of horses, dragons, insects, bears and birds, knights and maidens. A sample is shown in Plate 7.

In some visual alphabets, the letters are created from objects such as compasses, ladders or garden implements, but I find this a bit dull. How can a ladder possibly compare with a cavorting knight or fiery dragon for a memorable image?

One of the most famous writers about memory systems in the early Renaissance, Peter of Ravenna (c. 1448–1508), used alphabets from many languages to give him plenty of pegs. He claimed to have memorised huge amounts of information.

Peter of Ravenna, like many medieval people you'll meet, is called that because his surname is not recorded, just his place of birth. Peter used human figures to form the letters to create vivid, living entities, making his visual alphabet that much more memorable. He noted that his memory was best stimulated by images of enticing women, but warned against their use if the reader should loathe women or be unable to control themselves. I shall assume that my readers are neither sexist nor incapable of self-restraint and advise that the more vivid, stimulating and enticing your images of any gender, the more memorable you will make your information.

This penchant for obscenity was even more pronounced among the memory experts promoting the skills in the century before Peter. Despite being a theologian, chaplain to Edward III and briefly Archbishop of Canterbury, Thomas Bradwardine (c. 1300–1349) recommended obscene, violent or frivolous images. He deemed these shocking and therefore memorable images acceptable because a well-trained memory was essential for meditating on scripture and then preaching it. However, not all agreed. Saint Bernard advised the Benedictine monks of Cluny against such grotesque figures because distraction during meditation was a sin.

Thomas Bradwardine was a hugely influential writer on memory methods. He showed how puns could help to memorise the syllables of difficult words, something I use continually when learning the scientific family names of animals and plants, or new words in foreign languages. His ideas also had a major impact on my visual alphabet design. He encouraged the use of sequences of characters as memory devices, with each character interacting with the next in line. As an example, he showed a way to memorise the order of the zodiac, which was often used as a set of memory locations in the Middle Ages.

The first location of the zodiac is the central figure, Aries, a pure white ram, kicking an equally white bull (Taurus) to its right, who is retaliating in kind. Thanks to the kick, the bull has greatly swollen testicles, which are bleeding profusely. In front of Taurus are the twins (Gemini), being born in some grotesque way, either from a woman or the bull, one crying as he is pinched by a hideous crab (Cancer). The other twin caresses the crab: sibling love at work. Returning to Aries in the centre (the order dictated by medieval rules), the ram is also kicking a lion (Leo) in the head while it attacks a beautiful maiden (Virgo). Poor Virgo, despite her predicament, tries to balance scales (Libra) in her right hand while suffering from a dreadfully swollen bite from the scorpion (Scorpio) to her left. An archer (Sagittarius) has wounded a goat (Capricorn) to its right, which bleeds as it pours water from a jug (Aquarius) at its right foot for a fish (Pisces) at its left.

It is these interactions which give the sequence and make the zodiac more memorable. Consequently, my alphabet animals are far more dynamic than boring old sun–shoe–tree–door–hive–sticks–heaven–gate–sign–hen. That doesn't mean I never use the sun–shoe sequence; I do. I'll revisit it in Chapter 8.

My visual alphabet

We routinely use visual alphabets when teaching a child the letters. A children's visual alphabet is usually something like: A–apple, B–ball, C–cat, D–dog . . . Cats and dogs are capable of being active, violent, grotesque. (And may I add, cute?) But 'A is for apple' just doesn't do it for me. Each letter must instantly provide a memorable character who can become part of a lively story associated with whatever you're trying to remember.

The use of drawings ensures memorising the sequence is simple. Most people remember illustrations far better than they recall worded lists. So I drew my visual alphabet, as shown in Plates 1 to 6. Due to my love of spiders, I started with Arachne, who is throwing her web over the Bird of Paradise, who is being attacked by the Cat, who is being burned by the Dragon, who is also burning the Eagle who is not terribly impressed. The Eagle is flying in to attack the Frog, providing the link from Plate 1 to Plate 2. The Frog is standing on the horn of the Goat, who is being attacked by the Hydra, who is about to nip the Imp, who is kicking off the hat of the Jester. The Jester is going to spear the Kitten, who is blissfully playing with a piece of wool tied around the tooth of a roaring Lion. On the Lion's tail, a tiny Marmoset is being watched by a Neanderthal (with what intent we can only speculate). The ivy links the hiding Neanderthal to the Owl on the vine-covered tree stump, beneath which prowls a Panther, whose tail is wrapped by the tail of the feathered serpent Quetzalcoatl, who is attacking a curious Rat, whose tail threads the eye sockets of a Skull, which is being bitten by a Toucan, who rides the tail of a Unicorn. On the horn of the Unicorn, we discover on the next Plate, sits a Vulture, who stares at a Wombat, who is under attack from Xena the Wonder Woman. (Okay, I had to stretch the boundaries of medieval thought a little there, but it was a battle between Xena and Xerxes. As other people wanted to use my visual alphabet, I quizzed family and friends when deciding on the characters. Almost no one had heard of Xerxes, but Xena was familiar to most.) Xena is standing on a Yak, who is about to be zapped by Zeus.

By using interacting creatures, you will soon find that you know the order of the animals from their behaviour and not by the letter. The alphabet is always there as a backup if you need it, but just imagining the animals and their links makes recall much faster and smoother. And with ornate images, the viewer is required to pause to engage with the animals. A brief glance is less likely to produce indelible memories than a minute or two scrutinising the letters and the associated creatures.

Memorising anything, from a shopping list to a speech

Let's start with a shopping list. I want to remember to buy milk, tomatoes, the newspaper, a card for my granddaughter's birthday and a new watering can.

I imagine Arachne delicately throwing her web over the milk, but as she pulls it spills everywhere. She has also hooked the Bird of Paradise, as she always does. My Bird is particularly fond of his gloriously coloured tail-feathers, so I usually associate colour with him in some way. In this case, his tail is bright tomato red. The Bird is being stalked by a Cat, who has taken a toilet break on my newspaper. The Cat is in turn being attacked by a Dragon. The Dragon has also burned my granddaughter's birthday card, which I am not happy about—any emotional reaction will just make the image more memorable. An Eagle, furious that it too is being burned by the Dragon, puts out the fire with water from the watering can. Had the Eagle needed to remind me to buy a loaf of bread, he would have clobbered the Dragon with the loaf and eaten the toast. You'll always be able to find something to make the links.

I hope that you are now able to reconstruct my shopping list from the images created in your mind as you read that paragraph. The recall would be even better if you constructed your own images. This was brought home to me at a memory workshop. The tutor, Paul Allen, had created abstract sculptures from rusty iron and wood. The participants were asked to link them to the twenty largest countries of the world in the order of population size. James was very impressed by the associations made by other participants and used these in his set of memory images. When the time came for recall, James was able to name all the countries where he had made the association himself, but struggled with those he'd borrowed from others. Using your own imagined images is a critical part of all memory methods, including a visual alphabet.

I use my visual alphabet for public speaking. I love the way I can now make a much smoother presentation because I never have to glance at notes and find my place. You can do this too. Let's imagine that you are to give the speech at your brother's twenty-first birthday celebration and don't want to read it. To memorise it, you write out your speech in point form. You then associate each point with a creature from a visual alphabet.

Let's say that you want to open by acknowledging your parents. You imagine Arachne throwing her net over them. They struggle out to attend the party, being careful not to hurt the Bird of Paradise as they do so.

You then wish to make a joke about your brother's reckless motorcycle driving. For this, you imagine the poor Bird of Paradise clinging on for dear life as your brother speeds through the nearby streets, the bird's glorious tail streaming out behind. Next you want to be a little more serious and praise your brother's great computing skills. Imagine a cat constantly jumping onto the keyboard and driving him mad. Does he swear at it? Brush it aside? Tolerate its presence?

And so you continue. The interaction of the animals will ensure that you always get the next point without having to stop and think. As soon as you finish with Arachne you know that she always enmeshes the Bird of Paradise and immediately move on to that part of your speech. You'll find this method so much smoother and more natural than constantly looking down at notes.

I must confess that the first few times I gave a speech from memory, I was certain that I would forget something. As a safety, I had all the points neatly written down in various colours beside me at the podium, as I always had before. But I didn't glance at my notes once. When you speak this way, your hands are free, so you can hold up props or gesture to members of the audience. Should there be an interjection or sidetrack, you will find it very easy to recall the last animal you dealt with and continue, without having to stare at a bit of paper and find your place.

I often use my visual alphabet to store bird lists when out in the field with my husband, Damian, a very keen birder. It avoids the need to pause and juggle a notepad and pen alongside binoculars and camera. I simply write down the list when we get back to the car.

If the first bird we see is a magpie, Arachne gains a black and white dress, giving me the link. If the second is our next most common sighting, the Australian raven, then the Bird of Paradise will ridicule the raven's plain black clothes while preening his own multicoloured tail. And so the process goes on as we see each new species. The more you use your visual alphabet, the stronger their characters become and their growing personalities make it easier and easier to create the links. (I also act as Damian's living field guide, having committed to memory all 412 birds native to our state of Victoria, including their identification and other details, via a portable memory device that I'll describe in Chapter 5. I wouldn't attempt to remember that much information using a visual alphabet.)

You may choose to use my visual alphabet paintings, but if you create your own, the act of drawing your animals interacting, no matter how roughly, will quickly fix them in your memory.

Nearly two centuries after Bradwardine lived, the same memory methods were still being taught. In 1533, German Dominican Johannes Romberch, who wrote about memory training, suggested using the visual alphabets shown in Figure 1.1 for remembering names, among other things. While my natural memory had always been poor, my ability to recognise faces and remember names was truly appalling, so I tried Romberch's approach with great enthusiasm. I was sorely disappointed. Mary, Martin, Max, Michael, Mei-Lien, Madeline and Masahito, along with so many of their alliterates, all had to interact with my shy little Marmoset. I gave up.

But my desire to create a memory aid for names was reignited when I read about medieval bestiaries.

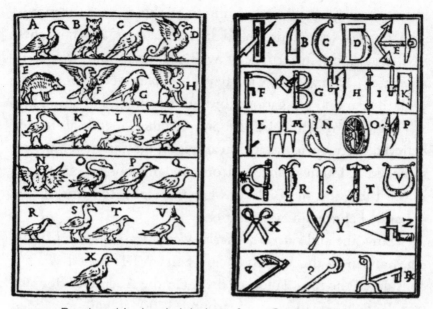

FIGURE 1.1 Romberch's visual alphabets from *Congestorium artificiose memoriae* (Venice, 1533).

Medieval bestiaries

The weird and wonderful bestiaries of the Middle Ages were books full of descriptions of both real and imagined animals. They were used to help readers memorise the multitude of moral expectations in that deeply religious time.

Medieval Christians believed that the characteristics of animals were designed by God to serve as examples of moral behaviour for humans. For example, the eagle was shown holding its young up to the sun and rejecting any eaglet who would not stare into the bright light just as, it was taught, God rejects sinners who cannot look directly into His divine light.

The heavily illustrated bestiaries served as a memory aid for the illiterate. They would remember the sermons associated with the images without needing to be able to read the words. The images were reproduced in monasteries and churches, in mosaics, on furniture, in wall paintings and woven in tapestries. Everywhere, medieval people were confronted with their moral obligations through the images of animals.

There is much debate about whether medieval people believed the many fantastical creatures actually existed. I suspect some did and some didn't. Unicorns and dragons are mentioned in the Bible and the Bible was taken literally, so some almost certainly believed in every critter.

Many fine and varied bestiaries were created in the Middle Ages and into the Renaissance. Henry VIII owned what is considered the most lavish of them all, the Aberdeen Bestiary (see Plate 8). Dating from around 1200, it has now been digitised and can be seen by anyone via an online search.

Bestiaries were often found in the libraries of monasteries in the Middle Ages. In *The Book of Memory: A study of memory in medieval culture,* medieval specialist Mary Carruthers wrote that scholars were often perplexed that such seemingly puerile texts were preserved in such lofty libraries. She argued that the bestiaries functioned not only as intriguing volumes presenting moral allegory in the form of the beast, but also as useful study aids. They provided an orderly set of memory hooks that could be used throughout the medieval student's education. They were a peg system.

Many artists have created bestiaries since. Even Leonardo da Vinci made his own. In 1899, the famous French artist Henri de Toulouse-Lautrec published one. I do feel rather pretentious giving it a go myself, but I'll survive that feeling. I wanted a reason to create a contemporary bestiary, so my failed visual alphabet for names arose, phoenix-like, from the ashes to become a contemporary bestiary for memorising names.

My bestiary for memorising names

One of the most common reasons people approach me for help with memory is to remember names. There are suggestions for this in most memory books. Among them is the advice to repeat the person's name, which works well but saying their name more than once or twice in a short conversation sounds very false. Another method is to try to link the name to something related, such as the surname Baker to a loaf of bread and Butcher to roast dinner, which might also work well except that I meet very few Bakers and Butchers and usually only want to remember a given name.

We're also told to associate the person with somebody famous of the same name, or someone else that you already know. Again, this works well if you can manage it. I am painfully slow at thinking of someone with the same name. I struggle to remember the names of celebrities at all, let alone to use them to help me remember someone else's name. Furthermore, in a society where it is fashionable to give unusual names and where many new acquaintances are from other countries, I soon gave up.

I simply couldn't quickly make a link without asking my new acquaintance to please shut up for five minutes while I thought it all through. I needed a totally reliable method that gave me an instant image to work with.

So I decided to create a bestiary based on the first two letters of a name. Instead of using just one letter of a visual alphabet for all 'M'-name people, I would be splitting them into 'Ma',

'Me', 'Mi', 'Mo', 'Mu' and 'My'-name people. Every animal has a significant feature: Aardvarks have fleshy noses, herons have elegant necks and macaws have heavy lines around their eyes. I would instantly know what animal to imagine during our initial conversation, and this meant the key feature I was looking at when I met a Mary or a Martin or a Max would be the same—the lines around their eyes.

Once I was sure I had memorised the first two letters, I could then reinforce the rest of the name at my leisure. My system would work for any name or surnames. In fact, the bestiary works as a memory aid for anything you can spell. I often use it to remember an unfamiliar word or the name of a café for a meeting.

I created a list of all possible word beginnings—'Aa', 'Ab', 'Ac', 'Ad' etc—which became disturbingly long, especially when the first letter was a vowel. I ended up with 264 pairs. If I couldn't find an animal starting with those two letters, I sought out a mythological beast, a plant or even (when desperate) an object. Medieval bestiaries were fairly relaxed in their definition of an animal, with mythological beasts and the real thing all merrily mixing, so I felt justified in stretching the concept just a little further.

'Ae' proved difficult but at least there were words to choose from; Aerialist had to do. 'Yv' was even worse. I could find no animal nor mythological character nor plant nor any useful noun starting with 'Yv'. I finally decided to design a scroll upon which was written the beginning of an apology: 'Dear Yv . . .', it began, and the rest of the name was obscured by the roll of the paper. 'I am so sorry.' My apologies to every Yvonne and Yvette.

'Mi' is millipede. When I met a Michelle or Michael, Millicent or Miranda, I checked out how well their eyelashes looked like millipede legs. Then I added something to help with the entire name. For example, Michelle became Mi-chelle; I added a shell to the eyelashes. To create an image I associated with her, I imagined Michelle with shell earrings. I could have imagined her with shell-like teeth, or tortoiseshell-framed glasses. I imagined Elizabeth regally riding her elephant. Philip arose from the phoenix's ashes, while Justin was linked to a juggler. In my imagination, he juggled balls and caught them 'just in' time. It really is good fun!

It didn't take me long to put most of the people I met into my system. As soon as the conversation was over (and I had repeated their name at least once) I mentally revised my image, reinforcing my memory of the name of my new acquaintance. Sometimes I couldn't immediately recall the animal I needed from the bestiary, especially when I started. I used my visual alphabet to remember the name's first letter and checked my written bestiary list as soon as I could. Before long, common names were committed to memory quickly because I knew the required animal and its feature instantly. Slowly, the entire bestiary became familiar.

My bestiary worked even better once I had sketched it. Maybe that was a result of the time spent sketching the animals and thinking about the initial letters. Maybe it was because I had stronger images. Probably both.

I became so enamoured with the idea of turning this sketch into my own bestiary artwork that I went to art classes and, with the help of the extremely adaptable artist Richard Baxter, I puddled around in watercolours, designed my own letters and learned to draw. (Plates 9 and 10 show an example of my bestiary, and the full bestiary is available from my website.[2])

There's a coffee shop near my art class. The first time I ventured in for a coffee to sip while drawing, I asked the names of the staff. Julie makes the coffee. 'Ju' is a juggler, so I imagined Julie juggling the coffee cups. Julie is my default female name for 'Ju', as most women I meet whose names start with 'Ju' are called Julie. The most common name I meet for any of the letter pairs becomes the default name. For example, Mary and David, Peter and Alison became the default names for Ma-, Da-, Pe- and Al-. If I meet someone with a default name, I add nothing more to my image. With Julie being the default, for Ju- I merely needed the image of my juggling barista to recall her name. 'An' is an angel so Angus needed wings, which I imagined tinted as black as an Angus bull. Ebony took my order then disappeared from sight. I hadn't been able to find any animal starting with 'Eb'. Instead, every time I thought about 'Eb', my brain responded with ebony and ivory, which I had rejected on the grounds that a piano isn't a critter. But I could find nothing better, and Ebony is the only 'Eb' name I come across anyway. So I imagined my coffee-order-taking Ebony was disappearing to tickle the ivories. The next week, every name came back with ease. (I take my coffee back to art class. Only once have I mistaken it for my water jar. I can confirm that paint does not improve the taste.)

I have found being totally honest works when I meet someone whose name is not familiar or easy to link. I say something like, 'I'm really bad at names and want to remember yours, so I have a system: I have to associate you somehow with a [millipede/piano/angel/juggler]. Do you have any ideas?' I may have to ask them how to spell their name. It works as a great conversation starter and cements the name for me. People are flattered that I think that remembering their name is so important. And it is.

Unlike the visual alphabet, the animals in the bestiary don't need to interact because it is not to be used in a strict sequence. Each pair of letters stands alone with its animal. But the layout of the page can help to make the creatures more memorable.

I don't think it's necessary for every reader to create their own bestiary, although taking the time to create 264 really rough sketches would probably work very well. My list of letter pairs and corresponding animals can be found in Appendix B. Of course, you will want to adapt some of them to animals that first spring to your mind; 'Gi' or 'Ph' gives me giraffe or phoenix, but you may well immediately think of gibbon or phascogale. Change them. Your first instinct will stick hard.

The bestiary and visual alphabet work together. If I meet someone whose name starts with two letters not in my 264, then I use the visual alphabet. Or if I can't remember my beast, I use the visual alphabet and at least get the first letter, and reinforce it later. I rarely find the first two letters insufficient to recall the name.

If I fail to remember a name when I next meet someone, I joke that my system has let me down and ask again. I've given up being embarrassed. The conversation will revert to the system and that will reinforce the name for next time.

I have developed the habit of doing a quick check through the day's names as I brush my teeth at night. If it is a name I really must remember, then I will jot it down and do another few revisions of my image and story over the next few days.

But mostly, imagining the link once upon meeting, once when the conversation is over, and once again that night serves my purpose beautifully.

Visual alphabets and bestiaries were just two of the memory devices used as writing became more widespread. I'll delve into the memory prompts built into the design of medieval manuscripts in Chapter 7, but for now we'll consider the most powerful memory technique of all. It has been around for tens of thousands of years and was taught throughout the Middle Ages and Renaissance. It's time to enter the memory palace.

CHAPTER 2

Creating a
memory palace

Why didn't I know about memory palaces decades ago? My natural memory really let me down at school; I struggled with the humanities because I simply couldn't remember one day what I had found absolutely fascinating the day before.

My poor mother. She loved everything to do with history and languages. While doing the dishes, she would quiz me on French vocabulary. Sitting in her extensive library, she would pull down book after beautiful book to show me gorgeous images of something she'd found really exciting about the Egyptian pyramids or the Incan marvel of Machu Picchu. Archaeology was her great love.

My mother was delighted when Damian and I both returned to university relatively recently, after long careers in education. I was researching the memory methods of indigenous cultures, but she was particularly pleased that Damian was studying archaeology. In 2008, we went to England. While Damian was off visiting every archaeological site he could reach, I had planned to explore the background for my thesis at the British Museum, my mother's favourite place in the entire world. I promised her that I would stay, as she always did, at the Bedford Hotel in a room that overlooked her beloved museum.

A month before the trip, my mother died. She never knew her influence would become so powerful; it was on that trip that I accompanied Damian to Stonehenge. It was there I first glimpsed the link between memory methods and ancient monuments, which became the topic of my previous book, *The Memory Code*. I am convinced that many of the monuments built by non-literate cultures functioned in part as memory palaces.

The first written record of memory palaces

The invention of memory palaces is often attributed to the ancient Greeks but they were just the first to write about them; memory palaces had been used by indigenous cultures around the world long before writing. Known as the 'method of loci', the 'art of memory', memory journeys or paths or trails, there is no more effective memory method known. All contemporary memory champions use it.

A memory palace is a visualisation of a trail of physical locations that are easy to remember in order. You can use a walk through your home, around familiar streets or any building that you know well. I have over a thousand locations in use as

I write and expect I'll have many more as you read this. I am shocked by that figure, yet I know memory champions who use many times that number.

The ancient Greek and Roman orators would create journeys through streetscapes or buildings consisting of hundreds, if not thousands, of locations. We know this from the writings of Cicero, Quintilian and the author of the amazing textbook the *Rhetorica ad Herennium*, which means simply *'Rhetoric: For Herennius'*. Gaius Herennius was some obscure guy about whom nothing is known but his name, who has nevertheless gone down in history because an anonymous teacher dedicated a textbook to him around 85 BCE.

The *Ad Herennium* advises students to choose locations very carefully, in a clearly defined sequence. The locations should be away from distractions, such as crowds of people passing by. They should be well lit, quite distinct from each other, of moderate size and with a moderate distance between them.

The ancient teacher suggested that each fifth place should be marked in some special way. For example, when using a building, I always make windows and doors the locations at multiples of five. So the fifth place might be a window and the tenth a door or doorway. This means that every group of five locations acts as a subset of the entire memory palace. When imagining the palace, I know that I must have four locations between each window and door. In the memory palaces through the streets around my home, I use side streets and corners as my markers for every fifth location. This rigour helps ensure that no location is missed and so no information is lost. I do this in all my memory palaces and find it invaluable. It sets up a rhythm as you mentally walk through your palaces and helps you recall every location and the information that you have stored there. Once the set of locations is firmly in place, you will find that you can start anywhere in the palace. Going forwards or backwards is no different in terms of mental effort.

You attach knowledge to the locations by imagining events there. The ancient text advises that reliable memory depends on creating visual impressions of incredible intensity. They need to be extraordinary in some way—striking, vivid, shocking—to make them highly memorable. The characters should either be extremely beautiful or really hideous. They should be dressed in strange clothes smeared with blood or paint or mud. They should be experiencing great tragedy or performing heroic deeds. They should be engaged in humorous situations or indulging in immoral behaviour or illegal acts. They may be encountering bizarre creatures or supernatural beings. And the more the images stir your emotions the better.

What the classical teacher didn't know is that these characters should be, in essence, just like the characters who had been populating the stories of indigenous cultures the world over for millennia. During the research for my PhD, I realised that indigenous memory trails are memory palaces that long predated Greco-Roman times.

Australian Aboriginal songlines

I learned most about memory trails from Australian Aboriginal cultures, of which there are at least 300 distinct language groups and more than twice as many dialects. Although these are distinct cultures, they all use sung sequences of locations through the landscape as memory palaces. In English, we call these songlines.

Songlines were used to help memorise everything Aboriginal people needed to know about animals, plants, navigation, genealogies, astronomy, the legal system, ethical expectations, obligations and responsibilities, weather, geology, seasons, land management, spiritual beliefs . . . the list goes on and on and on. Their culture was entirely stored in memory.

Indigenous Australians depended on their sung pathways through the landscape. We're not talking a stroll down to the river for an afternoon's fishing, we're talking massive distances. In his book *Singing Saltwater Country*, John Bradley refers to 800 kilometres of songlines known by the few remaining Elders of the Yanyuwa people of Carpentaria.[1] The whole of Australia was mapped by the local cultures who taught the paths to anyone with permission to pass through their territory. Can you imagine being able to recall every location along vast distances reliably enough to navigate?

But this is only the start of their astonishing achievements. At each one of the locations some form of ritual would be performed. A ritual is by definition just a repeated event. These rituals might be a song, a dance, a story or an entire ceremony but, critically, they encoded information. The performance might be about how to get water from the nearby water-hole, or how to knap the particular form of flint found at that location. It may relate to the people linked to that area or the animals and plants found there. Add together the thousands of songs and stories and dances linked to the songlines and you have an entire encyclopedia, fully indexed and sorted, ordered and made memorable. As one Elder explained to me, the song-lines act as a set of subheadings to an entire knowledge system.

It isn't only Australian Aboriginal cultures that use the landscape in this manner. Native American pilgrimage trails, Incan *ceques* and Pacific ceremonial roads all work the same way. It is a universal phenomenon because, as I'll show in the next chapter, this is the way the human brain works.

Creating a memory palace for the countries of the world

The easiest way to start is by creating a memory palace in your own home, or a location you know really well. Let's start with using your home to memorise the countries of the world, the first topic I tried. As soon as you understand the method, you can adapt it to any sort of information you want to remember, as long as you can list it in a defined sequence.

The first decision that needs to be made is the order in which to memorise the countries. Using alphabetical order would not be taking full advantage of the memory palace system, because the starting letter is already known, so I chose to memorise them in order of population, starting with China, the most populous, and ending with Vatican City, the least. I used the list from Wikipedia titled 'List of countries by population (United Nations)' which includes the 'inhabited dependent territories'; if you leave out those territories, you lose places like Puerto Rico and Guadeloupe. At the time of writing, there were 233 countries listed. The order sometimes changes slightly, depending on the most recent census, but I am perfectly happy to stick to my original list, which allows me to remember every country and a pretty accurate indication of the relative populations.

Here we go.

Start inside the front door of your home. We'll call this location one. Unimaginative, I know. The most populous country is China, so you need to somehow associate China with that location. You can either imagine a Chinese person knocking on the door, or a Chinese meal being delivered, or imagine yourself hurling a plate made of fine china at the door. Whatever you do, make the image active and colourful; nobody else knows what's going on in your imagination, so don't hold back. The title of Joshua Foer's bestselling book, *Moonwalking with Einstein*,[2] makes reference to the strange images he used to win the US memory championship. In a TED Talk, Foer said that 'the crazier, bizarrer, weirder, raunchier, funnier, stinkier the image, the easier it is to remember'. Foer knows what he is talking about.

You are going to put four more locations in this first room and then move on to the next room. This is the easiest way to follow the advice in the ancient textbook on memory, which recommended making every fifth location significant. Let's assume you put the second location at a bench, the third at a table, the fourth in that snug corner next to the bookcase and the fifth at the doorway to the next room. These are locations two, three, four and five.

At location two you need to make some association with India. The first thing you think of will usually be the most effective. I have an entire Bollywood production happening under the bookcase in the memory palace in my home. Location three, you associate with United States. Why not have Donald Trump sitting at your table? Depending on your politics, that image will lead you to curse or praise me forever because you won't be able to get rid of it. Location four you associate with Indonesia, location five with Brazil.

For me, Brazil is the window of the study where I am writing at this very moment. In my imagination, out my window is the glorious colour of the Carnival in Rio de Janeiro, the biggest festival in the world. A little man emerges from the noisy, dancing crowd waving his finger at me to remind me that Rio is not the capital of Brazil, it's Brasilia. The story of Brazil will just grow and grow as I acquire new information about it, but it will always be associated with that window, the Carnival and that little man.

Now we move on to the next room. Location six is a feature inside the doorway. You need to associate it with Pakistan. Location seven is the biggest country in Africa, Nigeria. And location eight is Bangladesh. Bangladesh? I checked the list thoroughly. Is Bangladesh really the eighth most populous country in the world? With more people than Russia and Germany and Japan? At the time of writing, yes. Why then doesn't it feature more in history? Why haven't I heard more about Bangladesh? Why hasn't Bangladesh used that massive population to take over other countries? When I first encountered this fact, I stopped memorising and started reading about Bangladesh. This happens constantly. You will very quickly discover that having a big picture and hooks for every piece of information just makes you want to know more.

Add country nine, Russia, and country ten, Japan, to your memory palace. In your mind, go for a quick walk around the two rooms and see if you have all the countries in place. Wait an hour, and see if they're still there. If you revise them only once or twice more, they will be there for good.

If you put every country of the world into a memory palace, then whatever you hear on the news, read in books or learn from friends can be hooked into those images, which are the starting points for stories to encode everything you want to know about the country. You can fill this memory palace for the rest of your life.

When the Caribbean was hit by hurricanes Irma, Jose and Maria in 2017, the news reports constantly referred to a string of mostly small countries. I knew almost nothing about them, but having those hooks in place made the news so much more real, giving life to the countries. Cape Verde, Cuba, Barbuda, Saint Barthélemy, Saint Martin, Anguilla, the Virgin Islands, Haiti, Puerto Rico . . . With each country's name, my mind instantly jumped to its location in my memory palace.

To know the geography of regions such as the Caribbean, I have created short songs giving me the countries in geographical order, and from that I can imagine the map. But I can't add more information—such as hurricanes and capitals, political situations or famous residents—into a finite song. Layering information is where a memory palace is simply unbeatable.

Because I have the memory palace my understanding of the world just grows naturally.

There's one lesson I didn't learn quickly enough: be very careful when encoding so that you get it right first time. It is amazing how quickly a good image will stick and how long it takes to get it unstuck again. When encoding the countries of the world down the main street of my hometown of Castlemaine, I put Samoa in the fruit shop with a lovely image of friendly Sam selling me bananas. But I had jumped countries: Samoa should have been in the butcher's shop four doors down.

Every time I saw the fruit shop, I would see Sam waving bananas. I had to add an image of Sam walking defiantly out of the fruit shop, down the street, and into his butcher's shop. It took a year of Sam stomping down the street before he stayed put, leaving my real fruiterer alone to drink his preferred liqueur which clearly linked him to the Caribbean island of Curaçao.

You can use a memory palace to memorise anything that can be organised: Oscar winners, horse race results, prime ministers and presidents, kings, queens and tyrants. You can store knowledge of plants and animals, mythical beasts and the rules for board games. If you can organise it, then a memory palace can store it.

Memory palaces in history

Five hundred years ago, readers were getting the same advice. You met Peter of Ravenna and Johannes Romberch and their visual alphabets in the previous chapter. In 1491, Peter published *Phoenix,* a book on the art of memory, which was in use consistently for the next 200 years. The number of locations in his memory palaces was just a tad more than in mine: Peter boasted that he used 100,000 memory locations, and I believe him. He advised that the best building to use for setting your memory locations is a quiet church. Peter would add new churches and monasteries to his repertoire whenever he travelled, circling their interiors three or four times to commit them to memory.

Johannes Romberch also described how to construct a memory palace. He drew an abbey where each of the buildings, such as the library, courtyard and chapel, served as a memory location, as shown in Figure 2.1. He drew a hand in every

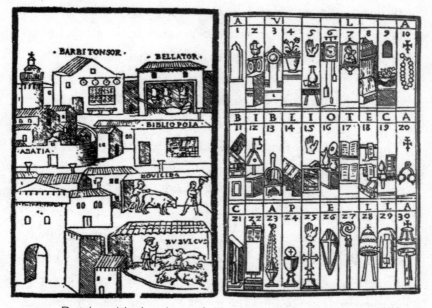

FIGURE 2.1 Romberch's drawings of the abbey as a memory palace from his *Congestorium artificiose memoriae* (Venice, 1533).

fifth location and a cross in every tenth, just as advised in the *Rhetorica ad Herennium*. If you draw your memory palace as Romberch did, you will find it much more memorable.

In 1966, British historian Frances Yates (1899–1981) published by far the most influential book on memory palaces of all, titled *The Art of Memory*. It had a massive impact on me; it was where I first encountered the history of the memory arts. She wrote:

The subject of this book will be unfamiliar to most readers. Few people know that the Greeks, who invented many arts, invented an art of memory which, like their other arts, was passed on to Rome whence it descended in the European tradition. This art seeks to memorise through a technique

41

of impressing 'places' and 'images' on memory. It has usually been classed as 'mnemotechnics', which in modern times seems a rather unimportant branch of human activity.[3]

A few pages later, Yates added:

There is no doubt that this method will work for anyone who is prepared to labour seriously at these mnemonic gymnastics. I have never attempted to do so myself but I have been told of a professor who used to amuse his students at parties by asking each of them to name an object; one of them noted down all the objects in the order in which they had been named. Later in the evening, the professor would cause general amazement by repeating the list of objects in the right order. He performed this little memory feat by placing the objects, as they were named, on the window sill, on the desk, on the wastepaper basket, and so on. Then, as Quintilian advises, he revisited those places in turn and demanded from them their deposits. He had never heard of the classical mnemonic but had discovered his technique quite independently.[4]

Unlike Yates, I am convinced that contemporary education would greatly benefit from using the ancient memory skills alongside literacy. I think my difference of opinion arises from two causes.

Firstly, Yates believed that the Greeks invented the method of loci and was therefore unfamiliar with the complexity and pragmatism of the memory methods as used by oral cultures. Secondly, she was not a practitioner and could only relate a trivial contemporary example. But it was Yates who brought the art of memory to the attention of the modern world and I shall be eternally grateful that she did. There are few books I rate as highly as hers.

Yates tells the story of Italian philosopher Giulio Camillo (c. 1480–1544) who decided that a purpose-designed building would work better as a memory palace than churches and streets, so he designed a memory theatre. It was to be able to represent all the knowledge of the entire universe; you can't fault his ambition. Camillo was already famous, so his memory theatre became a talking point right across Italy and France.

It was designed to take the form of a wooden amphitheatre, as shown in Figure 2.2. To use Camillo's theatre, you would stand in the middle of the stage and look up at seven semicircular tiers of seating. You would see arches and aisles, all decorated and divided according to the seven planets known at the time. There would be drawers and boxes in the endless rows, in which would be a printed version of everything already known, and (this is the bit I really want) everything that could ever

FIGURE 2.2 My impression of Camillo's Memory Theatre. (LYNNE KELLY)

be known. You just had to contemplate one of the mnemonic images for all the knowledge contained on the related cards to be called to mind. In this way, Camillo said, the spectator would be capable of discussing any subject as fluently as Cicero himself.

Camillo had designed a theatre in which, he claimed, the occult buried in the universe would be revealed, turning the method of loci into a mystical art. It was a pretty grand scheme, you have to admit. But was it ever built? There is a record that Camillo showed a wooden theatre in Venice and something similar later appeared in Paris, so perhaps it was. If so, nothing of the theatre remains. The details of how it actually worked were to be revealed only to the king of France, Francis I, Camillo's patron for seven years, but when the funds dried up in February 1544, he instead returned to Milan to draw up the final plans. Three months later he died, never having produced the great manuscript he had promised.

The modern tale of Solomon Shereshevsky and Alexander Luria

The story of Solomon Shereshevsky (1886–1958), however, is one we do know to be accurate. It provides the most remarkable example of an individual in recent times reinventing memory palaces with no knowledge of the long history of the method.

In 1929, the editor of a Moscow newspaper was annoyed with one of his reporters: at the morning meeting to allocate the daily assignments, one of his journalists hadn't bothered to take any notes. Unimpressed, the editor confronted the young Shereshevsky, who replied that he didn't need to take notes, he simply remembered what was said. The story goes that the editor picked up a newspaper, read a lengthy portion and challenged Shereshevsky to repeat it. Shereshevsky did so, word for word. The editor sent him for a mental examination.

Shereshevsky hadn't realised other people didn't memorise everything as he did. The confused young journalist went to the Academy of Communist Education and asked to see a memory expert. He was directed to the office of 27-year-old Alexander Luria, who went on to become a famous Russian neuropsychologist. Luria wrote a detailed account of Shereshevsky's extraordinary memory feats in his book *The Mind of a Mnemonist: A little book about a vast memory*, published in 1968.[5] Luria would present Shereshevsky with long lists of random numbers, mathematical formulae, letters, and even poetry in languages he didn't speak, which would all be recalled perfectly after a single hearing. Luria was never able to produce a list that Shereshevsky could not repeat perfectly.

Even more astoundingly, Shereshevsky could still repeat the lists decades later, as long as he was told where he had first memorised it. That location triggered the appropriate memory journey.

Shereshevsky possessed a formidable natural memory, but he also trained it. His memory palaces were the streets of Moscow. He converted words into vivid images, which he placed in well-lit locations along the streets. He talked about the need to ensure there was contrast in his images, and that the objects were sufficiently large. He avoided crowded streets, as they could confuse the clarity of the location. He also used all of his senses, linking aromas and textures and sounds to his extremely vivid, colourful and active images.

To learn foreign languages, Shereshevsky would reduce any word to the syllables and then memorise a sequence of images representing the syllables. Complicated mathematical formulae were memorised by creating a story linking each symbol within the formula. I'll return to these ideas in later chapters.

While he failed to make it as a journalist, Shereshevsky became a theatrical mnemonist, memorising increasingly difficult sets of data, often tables of random numbers. Although he was once given a table that anyone could remember—the first line was 1, 2, 3, 4, the second 2, 3, 4, 5, the third 3, 4, 5, 6, the fourth 4, 5, 6, 7 and so on—Shereshevsky did not notice the pattern; he used his mnemonic methods, as always.

I'm no longer as surprised by this as when I first read it. When I memorise a deck of cards, I use the method I'll describe in Chapter 10, which puts them in groups of three in memory palace locations. I've often noticed, when checking my recall later, that there were patterns: three eights or a nine–ten–jack sequence. I hadn't noticed the pattern; I just applied the method. But I suspect I would still have noticed lines of sequential numbers.

Virtual memory palaces

Some people prefer their memory palaces to be entirely imaginary. They create a building or landscape based on a favourite video game, movie or book. I rarely play video games and don't know any movies or books well enough to build a palace around them, but I know many people who use virtual memory palaces and assure me that they are very effective.

Modern as the concept of a virtual memory palace seems, there are many examples from medieval times. Theologian

Hugh of Saint-Victor (c. 1096–1141) constructed a mental image of Noah's Ark from descriptions in a book he knew very well, the Bible. Hugh had multiple locations within the huge ark in his imagination. He described the shape of each space, its colours and decorations. He intended his readers to imagine the ark for themselves, making it far more memorable. Hugh used his memory ark to recall the names of every generation from Adam to Christ. He then memorised the succession of apostles and popes, the evangelists, the geography of the Exodus, the zodiac, the seasons and weather, the 30 books of the Bible in order and, as was extremely popular at the time, lots of elaboration on vices and virtues.

You can populate a virtual memory palace with anything you like and, as Hugh did, add extra rooms wherever you need them. That's the big advantage of a virtual palace: you are not constrained by the reality of a physical space.

Continuous memory palaces—my History Journey

Everything I have said about memory palaces, and everything I have read, emphasises that they're based on discrete locations. It works superbly for countries: each country occupies its own position in the sequence. But what about history? Time does not come in discrete moments; it is a continuum.

I wanted to memorise Earth's history as well as its countries. (I want to know everything—I am not expecting to be finished any time soon.) And I wanted to start 4600 million years ago, at the beginning of geologic time. I also wanted to allow for individual years for the last century, so I could put in the births of family members and important historical figures. Clearly, I had a major problem. When I walk around the streets, I use a house block or shopfront for each location. If I allocated twenty metres for every year from millions of years ago, my memory palace would circumnavigate the world well over 2000 times. As it happens, I can walk through all of time in about two kilometres.

To begin, I split my History Journey into three sections. If I turn left at my front gate and walk around the block anticlockwise, I walk my Prehistory Journey, starting at 4600 million years ago and arriving back at my front gate at 1000 BCE. I've walked nearly a kilometre by then. If I walk around the block to the right, over a kilometre later I finish my History Journey at home in 1900. I then use my main memory palace for what I call my Twentieth-Century Journey, starting in my studio and stacking every year, from 1900 to the present, on top of the first 119 countries. The numbers in the memory palace match the dates: 1901 is in location one, 1902 in location two and so on.

The problem was setting up the scale for the first two journeys. The map is shown in Figure 2.3.

Let's start with the Prehistory Journey. Confusion arises fairly quickly between geological eons and eras and archaeological periods, so I set up a spreadsheet, as shown in Appendix C. The first divisions were assigned according to geological eons, starting at the front gate with the Hadean. As I cross the road through the Archean, the first house is the Proterozoic. Then I pass into the geologic eras. The next house, with its pale wall, represents the Paleozoic—its garden holds the first plants on Earth. The corner house with its very messy kerb is the Mesozoic, an easy association to make. I often find that saying a difficult word, in this case Mesozoic, a few times will lead

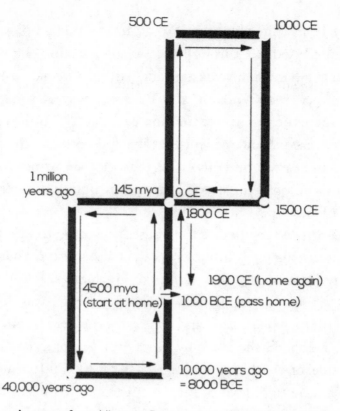

FIGURE 2.3 A map of my History Journey, starting from home at 4500 million years ago and arriving back home at 1900 CE. My Twentieth-Century Journey goes around the house and garden, with a location for every year.

me to make a link with the sound of a syllable. 'Messy' came quickly as I stared at the weeds.

The Mesozoic era is where we start the geologic periods and, even more exciting, get to greet the dinosaurs. They peer over the fence of the house at the Triassic era. I used to walk this Journey with my little dog, Epsilon-pi. Epsi would get scared at the start of the Triassic, but my rational brain refuses to believe it was because of imaginary dinosaurs.

Nevertheless, she would pull to go home unless I picked her up and carried her through the Jurassic, around the corner, which is 145 million years ago, and past the Cretaceous. Once we entered the Cenozoic era, Epsi was perfectly happy to walk again, despite the extinction event at 65.5 million years ago, so massive and violent that it wiped out the dinosaurs. She could never be convinced that the Mesozoic was safe and insisted on being carried through it right up to her death from old age.

I continued to allocate the houses according to the spreadsheet, reaching one million years ago by the next corner. I had already greeted the first hominids, including *Homo habilis*, *Homo ergaster* and *Homo erectus*, and knew that I was going to meet *Homo neanderthalis* and *Homo sapiens* very soon. From there, I divided the walk by years, not geologic names: the next side of the block was stretches of 100,000 years each. With a little rigging at the end, I turned the corner at 40,000 years ago. I parcelled the next side of the block into even smaller units so that I reached the last corner at 10,000 years ago. At this point I decided to change from quoting the date as so many thousands of years ago to using BCE, so I had to take off 2000 years: as I turned the corner, my Prehistory Journey jumped from 10,000 years ago to 8000 BCE. After passing a lot of the archaeological sites I wrote about in *The Memory Code*, I arrive back home at 1000 BCE.

As I gather each new bit of information about the prehistoric era, I add it to the spreadsheet, which is getting longer and longer. But my walk around the block remains the same length: there is always something—a fence post or a tree or a feature of a house—on which to pin a new association.

Why absolutely everywhere needs a name

I mean everywhere. And I mean a memorable, useful name. When setting up a memory palace, name every location carefully and it will serve you well. That is a lesson I had not learned well enough when I set up my first four big memory palaces, five years before writing this book: Countries Journey, Prehistory Journey, History Journey and Twentieth-Century Journey.

In the Countries Journey, every location had a name by default: the name of the country. When I announced to Damian that 'Kirsten bought Tokelau', he knew that she was moving in around the corner and had not, on a schoolteacher's salary, suddenly bought an entire island.

But my Prehistory Journey posed a problem: I couldn't instantly recall the dates of most of the locations that I had chosen as the key divisions. I knew where the locations were, but I had noted them in the spreadsheet as 'edge of the fence' or 'blue house'. Not exactly memorable.

Why did it take me so long to realise the problem? It's not as if I didn't have plenty of indigenous advice about the importance of naming locations; Aboriginal Australians and Native Americans had long emphasised the way they sing the names of locations to recall the pathways through their landscape memory palaces.

The first location in my Prehistory Journey is the stone wall in front of our house, which I had referred to as 'our stone wall' and linked to the Hadean geologic eon. The Hadean? A wall? And it hadn't occurred to me to call it Hadean's Wall? I hang my head in shame.

The driveway at 65.5 million years ago, when the dinosaurs were wiped out, is no longer referred to as 'long driveway' but

'Dead Dino Drive'. At 100,000 years ago, the house that I had listed by the name on the gate (which I kept forgetting) is now 'Cent Mansion'. The power pole, previously known as 'power pole', is now 'Thirty Dirty Chauvet Pole', which rolls off the tongue rather nicely, don't you think? It also tells me that I am at 30,000 years ago (the rhyme helps enormously), when the Chauvet Cave was so wonderfully painted. The little gap at the bottom of the pole (a cave, if you have a vivid enough imagination) confirms that I have the right one.

After two walks through my Prehistory Journey saying the new location names aloud (once I had checked there was no one within hearing), they were all safely in place.

My History Journey

At the end of the Prehistory Journey, at 1000 BCE, I can drop in at home for a cup of tea or continue on into the History Journey. I chose 1000 BCE more for my convenience than anything else. The division between Prehistory and History has been defined as the time when written records started, but this date varies all over the world. For my History Journey, I walk from home to the corner of the next block, which is 0, debating with myself whether I'm at zero BCE or CE. When I set up the History Journey, I started by labelling the corners of the block: 0, 500, 1000, 1500 and then back to the first corner, which I designated 1800 CE because I need extra room for more recent events. The walk from the corner to my house gets me to 1900, where it joins onto the Twentieth-Century Journey.

I then subdivided each side of the History Journey block, finding some significant location to mark each 50 years—the start of a house block or side road, bus stop or tree. After 1400 CE,

I subdivided further, giving me 25-year jumps between locations from then onwards.

Let me describe what it is like to walk my History Journey.

From my front gate to the corner of my block takes me to zero. In just this short walk, I have acknowledged Homer, passed through the Chinese Zhou, Jin and Han dynasties, seen Romulus kill Remus, and glimpsed the tumultuous lives of Nebuchadnezzar I, Pythagoras, Confucius, Buddha, Herodotus, Socrates, Plato, Alexander the Great, Euclid and Archimedes, Cicero, Julius Caesar, Antony and Cleopatra, and Christ. Work on the Great Wall of China has begun and the Silk Road is carrying traders east and west.

Every side of the block is crammed with living history. I can see what will happen ahead and turn to see what has led to the event. Around me, at any location, is what is going on all around the world at that time.

As I step from BCE to CE, the Dark Ages loom in the shadow of the Catholic church hall down the road. Barbaric hordes descend the downpipe to sack Rome in 410 CE. As I enter the building's shadow, Augustine dies at Hippo (the bulge in the pipe) and Attila does his Hun thing into the Balkans. Turning the 500 CE corner, I can see Muhammad's birth just ahead, and the Sui Dynasty about to unify China. A little while later, escaping the smell of King Alfred's burning cakes, I round the corner at 1000 CE. I pass the flourishing of Great Zimbabwe in Africa, and pause in the grounds of the local school, at 1431, where Joan of Arc burns. I smile as I remember how little Epsi often stopped here to pee at the martyr's feet. I had refrained from explaining to her that the tiny output was pitifully late to be of use.

Around the corner, at 1500 CE, the young Henry, born just behind me, is taking the throne as Henry VIII. There's Kepler and Galileo and Pascal and Watt and so many more of my heroes from my days of teaching physics. At the last corner of the block, at 1800 CE, I turn for home.

From this corner I need to retrace my steps, so I cross to the other side of the road. Although Socrates lines up pretty well with Queen Victoria, they have a road and 2000 years between them.

Between every event and every person there are shrubs and letterboxes and garden beds aplenty to add events. If I want to remember a specific date, I use the Dominic System for numbers, which I'll explain in Chapter 4. Mostly I just want to know what was going on around the same time, and approximately when things happened.

I add events or people whenever I want to. I can add a new event at the appropriate date at any point on the Journey. At the location when Joan of Arc dies, there was a woodpile, which made the association rather easy. Even though the school later replaced the wood with a water tank, the woodpile is fixed in my memory and Joan of Arc stays put. Later, I wanted to add Mehmed the Conquerer, born in 1432. Wonderful. I now see the tank falling from the heavens and smashing poor dear Joan to pieces as the water puts out her fire. Mehmed had conquered her.

The Pueblo Revolt is in 1680. There is a PRIVATE, KEEP OUT sign on the fence. Perfect. The Pueblo were keeping out the Spanish. The sign is between 1675 and 1700. It is nearer the former, so when I think of the date, I know that it is nearer 1675 than 1700. I will be very close to the actual date simply by knowing the location. Just a bit earlier, at 1644, is the start of the Qing Dynasty in China, which was ruled by the Manchus. So I have placed a picnic on the ground there: a *man chews* on his food and calls (with a Qing sound) for more. I can look ahead and see that the Qing Dynasty is going to continue right around the corner and into my garden, when the Chinese Republic supersedes it in 1912.

I might walk the block just to focus on information about the British kings and queens, say, or the Chinese dynasties. Or I might pause at a location, perhaps 1200, and see King John in England while Great Zimbabwe is flourishing in Africa. I also like to ponder at one spot, looking forward and back in time. I am forever playing with the journey mentally. I never do the whole journey recalling every single event and person. It would take too long.

I use the History Journey almost daily. When any historical event or person is mentioned, my brain will automatically jump to the appropriate location. Before I had my History Journey, I had never noticed how often historic events and people are referred to in everyday life.

One day, Damian was musing about the relationship between scientists, artists and musicians as we walked through the seventeenth century. 'What I like about this memory palace stuff is all the connections—both deepening and broadening of knowledge.'

I was impressed. 'That's profound.'

'Of course. I don't say much, but what I do say is profound.'

As I walk these streets—heading to the shops, the library or my art class—historic characters keep clamouring for attention. There are times I just tell them to shut up. *Can I please have a quiet walk here in the present?* Usually they oblige, but Saint Augustine can be very persistent. Maybe it is because of what he said about memory palaces.

When still the mere mortal Augustine of Hippo in the fourth century, he wrote:

> And I come to the fields and spacious palaces of my memory, where are the treasures of innumerable images, brought into it from things of all sorts perceived by the senses . . . When I enter there, I require what I will to be brought forth, and something instantly comes; others must be longer sought after, which are fetched, as it were, out of some inner receptacle; others rush out in troops, and while one thing is desired and required, they start forth, as who should say, 'Is it perchance I?' These I drive away with the hand of my heart, from the face of my remembrance; until what I wish for be unveiled, and appear in sight, out of its secret place. Other things come up readily, in unbroken order, as they are called for; those in front making way for the following; and as they make way, they are hidden from sight, ready to come when I will. All which takes place when I repeat a thing by heart.

Mnemonic verses

There are many mnemonic aids that seem a lot simpler than setting up a memory palace. For example, there is a rhyme for the monarchs of England since William the Conqueror, which can be sung to the tune of 'Good King Wenceslas':

Willie, Willie, Harry, Stee,
Harry, Dick, John, Harry three;
One, two, three Neds, Richard two,
Harrys four, five, six . . . then who?
Edwards four, five, Dick the bad,
Harrys twain, and Ned the Lad;
Mary, Bessie, James the Vain,
Charlie, Charlie, James again . . .
William and Mary, Anna Gloria,
Four Georges, William and Victoria;
Edward seven next, and then
George the fifth in 1910;
Ned the eighth soon abdicated
Then George the sixth was coronated;
After which Elizabeth,
And that's the end until her death.

The trouble with rhymes like this is that you can't add any more information. What if I want to know when Queen Victoria came to the throne, the name of her husband and children, or other events of her life? In my History Journey, Queen Victoria is linked to a large rock embedded in the cliff along the edge of my road. There are plenty of crevices, bumps and lumps all over the rock to which I can hook on an endless array of facts about Queen Victoria. I can see what leads up to Victoria's rock and what else is going on in the world at the same time. I am limited only by the time I have to devote to it.

A memory palace is a structure, grounded in the landscape, offering a firm base on which to build a tower of knowledge to play with, analyse and think about—a way to ponder the big picture.

Dominic's Rule of Five

I'm often asked how long it takes to mentally scan a memory palace and locate a specific detail from a location. The answer is recall is much faster than any other method because you are thinking in images not words; your brain can scan images extremely quickly. You don't need to imagine yourself trotting right around the palace; the keyword will jump you to the right location. But you do have to make sure the information sticks when you encode it. Here's some great advice about how to do that from the world's best-known name in memory sports.

Eight times world-memory champion Dominic O'Brien started memory training because he struggled with a poor memory. He claims no special natural ability, yet he once memorised 54 decks of cards after just a single sighting of each card. He is the author of fourteen bestselling books on memory, has won many awards and regularly appeared on radio and TV in his native United Kingdom and around the world.

Like all memory champions that I'm aware of, O'Brien uses memory palaces. He coined the term 'the journey method' for the way you move through a memory palace, collecting information from each location in order as you go. The term is now widely used. It is a perfect description of the way I use my History and Country journeys, both of which I refer to almost daily.

It was a huge privilege to spend time with Dominic when I was in the United Kingdom in 2017. After offering me all the guidance I could cope with at that early stage of training, he has been drip-feeding me advice ever since and become both a friend and mentor. I'll return to memory competition training in Chapter 10, but Dominic's influence on my thinking extends far beyond the competition hall.

One of Dominic's most valuable contributions to the world of memory is his Rule of Five. It states that we should recall information strategically by using the following pattern:

First review: Immediately
Second review: 24 hours later
Third review: One week later
Fourth review: One month later
Fifth review: Three months later

The Ebbinghaus Forgetting Curve represents research which shows that after an hour we forget more than half of what we have learned. A week later, we are down to about 20 per cent retention. It is repetition, such as Dominic suggests, that transfers knowledge from short-term to long-term memory.

With a memory palace, I have found those five revisions ample for pretty reliable permanent memory. Without a memory palace, even more than five repetitions would not work for me at all.

I had not revised my Countries Journey for well over a year when I started writing this book. Some of the countries had been in the news, or had been mentioned by friends during that year, and my brain had jumped their locations, but well over half of them had gone unremarked. I could still mentally walk the entire memory palace with ease, although a dozen or so of the capital cities needed a quick check from my list.

I would probably recommend an annual revision, just a quick mental pilgrimage to each palace, to ensure all is in place.

Dominic O'Brien developed the memory palace technique independently before he learned that the Greeks and all indigenous cultures had beaten him to it. The fact that people create the concept of memory palaces independently shows just how much they are a natural way of thinking. Every culture I have explored in the last decade of research has used memory palaces in some form. They didn't all chatter with each other across thousands of miles and millennia of time. There must be a reason that they each developed such similar memory techniques. That reason lies in the human brain. It's time to look at the neuroscience of memory.

CHAPTER 3

Stories, imagination and the way your brain works

We have a great deal to learn from the way ancient indigenous cultures memorised all the knowledge on which their survival, physically and culturally, depended. We also have a great deal to learn from the most recent developments in neuroscience. Combining the lessons from these diverse sources can give us an unparalleled insight into our own memory capabilities.

The most common application I was asked about by readers of *The Memory Code* was how memory systems could be used in learning foreign languages. Why did it have to be foreign languages? I have always found learning them impossibly difficult. At school, I tried French, Latin and German—I was appallingly bad at all and enjoyed none. But I couldn't escape the fact that readers wanted to know about learning foreign languages.

I resolved to try again with French, a decision made from a sense of duty rather than any pleasure. And if I was going to suffer for my readers, I felt I should do it thoroughly, so I added Chinese: a language in which nothing is familiar, seemingly written with indecipherable squiggles.

A year later, I have changed my opinion totally. I am loving learning languages.

But first, the lessons from indigenous cultures and neuroscience.

As I began my PhD research, I was soon seriously impressed by the extent and depth of the knowledge encoded in the narratives of oral cultures the world over. And when I learned more, I was absolutely staggered.

Indigenous knowledge systems

I quickly stumbled across a reference to a study of the Native American Navajo, which found that the Navajo had classified over 700 insects and stored the entire classification in memory. I had no doubt this was exaggeration, but I simply had to get that study to be sure. There was not, as far as I could detect, a single copy in Australia. The academic librarians at La Trobe University did those impressive things that only university librarians know how to do and tracked down a copy in the United States. Three months later, I had an original copy of *Navaho Indian Ethnoentomology* by Leyland C. Wyman and Flora L. Bailey, published in 1964. I suspect this particular copy had not seen the light of day since—the musty smell was almost overpowering.

But there it was: page after page, 701 insects, each with their Navajo name—genus and species—and a bit of information about each. The Western classifications followed, which were remarkably similar. Wyman and Bailey also pointed out that the Navajo made these classifications because they sought knowledge for knowledge's sake: only a few species were noted because they are 'botherers'—lice, sheep ticks, flies, gnats, mosquitoes or pesky ants—or eaten, the cicada. The authors commented on the remarkably accurate observations of insect behaviour and their habitats. They also referred to the way the information is stored—in mythology, in sand paintings, and as metaphors for many aspects of the culture. (A long list of Western entomologists were thanked by name for their contributions. The unnamed indigenous 'informants' were acknowledged as a group, none as individuals.)

And that's just insects. Mention was made of the Navajo's extensive knowledge of birds, plants and every other aspect of the environment. And it was all stored in memory.

Studies like this are very hard to find because they require a whole team of scientists, linguists and indigenous experts working together for a very long time, and in incredible detail, across cultural barriers. I gradually managed to find examples from cultures across the world. The Hanunóo in the Philippines classified 1625 plants, many more than known by the Western scientists in the team. The Matsés peoples of Brazil and Peru recently recorded their traditional medicine in a 500-page encyclopedia, all from memory.[1] For months, I was enthralled by research about the Pacific navigators and the vast amount of knowledge and training they needed to cross thousands of miles of open ocean. Then there are the incredibly complicated scientific knowledge and lengthy genealogies of the Māori of New Zealand.

I was influenced most of all by studies with Aboriginal Australians, detailing incredible knowledge of every aspect of their environment.

Across all the studies certain elements kept appearing. In every case, there was mention of the way the knowledge was encoded in mythology, in vivid stories that not only told about the animal or plant but also integrated the knowledge with what might seem to Western readers as dissimilar aspects of their culture. Indigenous learning is not divided into discrete silos separating science from history, spirituality from navigation. It is all enmeshed in an intellectual whole.

All of the cultures used song, or song-poetry, because song is far more memorable than straight prose. And they all danced. They sang and danced vivid stories of their practical knowledge and swags of cultural information as well.

Music is a powerful memory aid. How much better do you remember songs from your youth than the things you read? Then there's that incredibly annoying tune you can't get out of your head. So why not make that tune about something you *want* to remember and welcome its presence?

Vocal teacher Christopher Lincoln Bogg described song to me as 'the enhancement of speech in an emotionally charged framework':

Song follows the intonations and rhythms of spoken language.

We use our voices together for a reason. Song unites us in our communities, at church services and football games. It forms the emotional map of our oral histories. Song is for most of us an agency for emotional release. Unlike any instrument, the human voice is hard-wired directly to our emotions. Sing, and the emotions feed back to you. Listen to a good singer, and you follow the emotional ride.

Indigenous people from a whole range of cultures talk about the power of the emotion in their songs. I expect that, like me, you remember the songs you connected with emotionally far better than intellectual lyrics. That is why their sung mythology is so full of exciting events.

'Mythology' in an indigenous context is not like the word 'myth' in English, which implies something untrue. English doesn't have a word for 'vivid stories that encode rational and practical and spiritual information of which the literal truth is impossible to differentiate from the metaphorical, a dichotomy not usually relevant to the traditional owners'. So we use the word 'mythology'.

I live on Dja Dja Wurrung country. Local Elders asked that I call this combination of stories, dance, mythology and lessons for life 'teachings', and they ask that for very good reasons. Their entire knowledge system is bound up in those stories and songs, taught through performance. If the songs and stories are entertaining, that is good. If they stir the emotions, that is better. And if they educate, that is best of all. So, sing and dance and tell stories about the teachings you want to remember. You will be amazed by the difference it makes.

Indigenous teachings are often embellished with glorious artworks, narrative flourishes and elaborate dances in magnificent costumes and masks, but the core teachings always remain intact. I have found a huge range in the art forms which enhance the knowledge system, but that's for Chapter 5.

What I learned from indigenous cultures was that if I wanted to use my memory effectively, I should incorporate song, dance, mythological stories and wildly emotive images. And, of course, there was the universal implementation of memory palaces.

Memory and the human brain

I have always thought of cells as very small things, microscopic even. They are, in my imagination, tiny little round things with a dark, coloured-in nucleus, just as I drew in my science book at school. In fact, I should have discovered decades ago that some of our cells can be over a metre long; these are neurons—specialised cells that make up the nervous system and transmit information all around our bodies. They come in a vast array of shapes and sizes. Although the neuroscience of memory is extremely complex and far from fully understood, it's clear the hippocampus is the area of the brain that plays a vital role in learning and memory. Critically, it consolidates short-term memories into long-term memories. This was discovered in 1953 when a patient, known in academic literature as HM, had his hippocampus removed to cure his severe epilepsy. Although he was still able to access long-term memories from other parts of his brain, without a hippocampus he couldn't store any new memories. This disability was severe enough to mean he could no longer live independently.

Dr Jenny Rodger is Associate Professor in Experimental and Regenerative Neurosciences at the University of Western Australia. She specialises in studying the cellular and molecular mechanisms of brain plasticity in relation to brain development, learning and memory throughout life. Dr Rodger described the hippocampus to me as 'the main seat of memory'. She explained that the brain is particularly good at associating memories with place. Apparently, the hippocampus gets especially active when learning spatial knowledge, so it is logical that any knowledge encoded by using positions in space will lead to greater activity in the hippocampus.

If you deliberately make associations, such as considering a location within your memory palace as you absorb new information, the two concepts will become linked in your neural pathways. Concentrating on two cognitive images at the same time ensures that one will trigger the other. This is why when I say the name of a historical character, I get the image of the location on my History Journey. If I think about a location, I will be able to see the historical character I mentally put there. I am getting what neuroscientists call a 'temporal snapshot'.

Dr Rodger explained that in order to maximise neurogenesis—that is, the creation of new neurons from neural stem cells—you want to be thinking new and creative ideas. Novelty is a great trigger for neurogenesis.

This matches my experience with memory methods. The more vivid and wild and unusual and grotesque you make your images associated with the memory palace, the more likely the process is to succeed. Mundane information won't excite your brain, but characters and stories and wild ideas will.

Dr Rodger also explained that the moment you learn something neural circuits are physically altered in your brain. I was delighted to learn that recent research is debunking the old idea that the growth of our brains is somehow finalised in youth and that as we age we only lose neurons and fade into vagueness and confusion. It turns out that we are capable of adding to the complex circuitry within our brains *throughout* life. It's a matter of keeping on using it, rather than settling into cognitive inactivity and leaving all the hard thinking to the young. This reassuring research comes under the general topic of 'synaptic plasticity'. This term comes about because of the research which demonstrates that your brain's circuitry is not just about the neurons themselves but also relies on the gaps where the neurons interact, the synapses. Neurons communicate chemically across the synapses, creating the neural networks which store memory.

Your brain will constantly strengthen the synapses that are most active. The chemistry, physics and biology which underpin the reason repetition strengthens synapses and makes memories hang around long term is really interesting. Unfortunately, it is beyond my scope here. (For those with the inclination, I suggest typing 'NMDA receptors' into your search engine and spending a few hours among the ions and proteins and glutamates which are fundamental to the way our memories work.)

Your brain also uses associations to form links. So concentrating regularly on certain knowledge will strengthen the appropriate synapses; the more you associate new ideas with familiar knowledge, the stronger those ideas become.

But it gets even better. I was astonished to discover that the places that we know well—our homes, offices, schools—are recorded in our brains as actual physical pathways made of neurons. Unsurprisingly, the better you know your memory palace, the more robustly it will be represented in the physical structure of your brain.

It is no surprise that indigenous cultures all over the world independently created memory systems based on the physical landscape, skyscapes and decorated spaces of all sizes—hippocampal place cells are really good at representing the spaces you encounter and your place in them.

If you want to head off into the virtual world again, your key term for searching is 'entorhinal grid cells'. It is all very recent research and earned the scientists the 2014 Nobel Prize in Physiology or Medicine. Edvard Moser, May–Britt Moser and John O'Keefe showed that these cells constitute a positioning system in the brain. They encode a cognitive representation of the physical space you inhabit.

So memories that piggyback on the hippocampus's ability to remember space can become very strong: any events we encounter, or associate with, specific locations in space at a given time can gain this potency. After our discussions about the memory techniques I had been researching, I was delighted when Dr Rodger wrote: 'It is satisfying to see this very ancient neurological association of spatial information and memory processing within the hippocampus applied consciously in the memory techniques that you describe.'

(Oh, and Dr Rodger mentioned something else: goldfish can navigate mazes they have learned for weeks afterwards. The 'fact' that they only have three-minute memories is rubbish.)

Exceptional memorisers are made, not born
Swedish psychologist K. Anders Ericsson is best known for his research on expertise and human performance. In one of his many experiments, on experts across an impressive range of fields, Ericsson concluded that it is hours of practice that generates experts: natural ability helps, but persistence is the essential ingredient.

In this study, Ericsson explored the brain function of memory champions. His question was whether exceptional memorisers develop that skill or are born with some kind of mental advantage. (But let's dispel another myth before we go any further: Ericsson is convinced that there is no scientific evidence that anyone has a 'photographic memory'. If so, one of these individuals would be sweeping the floor with memory championships, but there are none competing. There are plenty of people with extremely good natural memories, but this would not rank them in the league of memory champions, as will become clear when I explain the competitions in Chapter 10.) Ericsson found that most memory champions started training because they had *poor* memories, but even those who were already strong scholars had to train a great deal in order to compete at a world-class level.

Along with his colleagues, Ericsson compared images of the brains of a group of world-class memory champions to those of a control group of a similar age and background. Memory champions memorise lists of digits, unrelated words and the order of shuffled decks of cards. Ericsson's research showed that memory performance for abstract data dramatically improved with training for all participants. They found no systematic anatomical differences in the MRI images for the brains of memory experts and control subjects. All but one of the memory performers said that they used 'the method of loci', while none of the control group used standard mnemonic techniques. Unsurprisingly, the scans showed that memory experts had higher activity in the areas of the brain now known to be linked to spatial memory and navigation. Ericsson concluded that the research provided compelling evidence that dramatic improvements in memory are possible for ordinary people willing to learn and practise memory techniques.

But Ericsson's research did not extend beyond the techniques used by memory champions; significantly, it did not include using an integrated form of memory methods as is typical in indigenous cultures. Different techniques suit different types of information.

In a similar experiment reported in 2017, neuroscientist Martin Dresler at Radboud University in the Netherlands, along with Boris Nikolai Konrad, a memory coach and athlete, recruited 23 of the top 50 memory competitors in the world. Along with colleagues at Stanford University, they also found that the brains of the memory athletes showed no structural difference to the control subjects: it was training that altered the brain's neural networks, increasing memory performance. This exactly matched what the memory champions themselves had reported; their training had given them the advantage.

Putting it all together—learning foreign languages

Emboldened by this research, I resolved to take all these methods—song, dance, characters, memory palaces, imagination—and apply them to learning foreign languages. I had been forced to study a foreign language until the end of Year 11, had settled on French, hated it and barely passed every year. My final pass was granted on the condition I didn't continue with French, which was fine by me. I remember almost nothing from those five years of lessons all those decades ago.

And starting a foreign language in my sixties was certain to fail, wasn't it? I had often heard that young children are the only ones who can learn languages well. Perhaps this was an out: maybe I was too old?

Dr Barry McLaughlin from the University of California is an expert in learning a second language. He argues that the widely held belief that young children learn languages much better than adults or adolescents is nonsense. His research shows that, given the same number of hours of exposure, adults will learn far faster than young children. Unsurprisingly, young children use simpler grammar, talk about simpler concepts and use a much smaller vocabulary. Even in language-immersion schools, children who begin language studies in secondary school perform just as well on final language tests as those who start at the beginning of primary school.

There is an exception: some studies, although not all, indicate that children learn pronunciation better than adults. Young children are considered to be far better mimics, a skill we do lose as we age.

There went my excuse. I had to learn a foreign language. The obvious choice was French, as I had studied it before and it sounds so lovely.

Learning French

In my first week of trying to learn French I decided to use two characters to help me with masculine and feminine. One thing I did remember from school was how difficult I found the noun genders. Is a fork masculine or feminine?

Indigenous cultures the world over link stories to a pantheon of mythological and biological ancestors. As I shall describe in the next chapter, using characters with distinct personalities makes everything more memorable. So I employed a pair of teddy bears in my French lessons. The large Fleur is decidedly feminine, her crimson dress embroidered with flowers. The much smaller male bear, Le Petit Professeur, has deep purple fur and is dressed in nothing but a ribbon. The differences in size and dress make the images distinct and therefore more memorable (see Figure 3.1).

FIGURE 3.1 Le Petit Professeur and Fleur, with objects sorted according to their gender in French. (LYNNE KELLY)

The bears and I started with clothes as I got dressed each morning. If the item of clothing is feminine in French, then Fleur would hold it, chat about it, engage with it and make it memorable. She wore my glasses and was wrapped in my skirt. Petit Prof was rather uncomfortable about wearing a bra, but in French *le soutien-gorge* is masculine so he had no choice. To remember the word, I imagined the bra 'suiting a gorge', there being no cleavage on his little furry front. However, despite the French language demanding the word was masculine, Petit Prof outright refused to have a vagina. After a few attempts to recall the word using the memory aid, it just became part of my vocabulary and the silly link was no longer needed.

I tried this method with a French teacher and her young students in a primary school. (You'll be relieved to learn that the topic was food: no bras, just baguettes.) It worked a treat. The children made no mistakes, easily remembering if the words were masculine or feminine. They also fell in love with Fleur and Petit Prof.

The presence of two French-speaking bears also prompted me to chatter. Without the bears, I would not have discussed what I was doing aloud. Repetition is absolutely essential for memory. No human would have put up with my intolerably slow and extremely repetitive utterances, constantly punctuated by dashes to dictionaries. My vocabulary grew fairly rapidly for the items that I use every day.

I discovered that two local bookshop staff speak French. I had found a captive audience of real humans to practise on. At that very early stage, I had only conversed with my bears about clothes. So unless Wendy and Kathryn were particularly interested in how I got dressed in the morning, I had no conversation. I started practising light chatter. Fleur, Petit Prof and I became obsessed with greeting each other, commenting on the weather and asking each other how we felt today. The bears require me to use all the personal pronouns: 'I', 'you', 'we', 'he' and 'she', and I use 'they' when Damian wanders by. I tell him what the bears have been doing. He just says *'oui'* and keeps going.

Fleur and Petit Prof started to help with the cooking, gardening and any other not-very-brain-engaged activities, and I gradually built up vocabulary lists. With my tablet at hand, I would look up phrases that I needed with online translators, which conveniently provide pronunciation.

Online translators aren't an entirely reliable way to go. After a few months, I discovered that my bears and I had been baking cakes in the rubbish bin, when I thought I had been saying 'cake tin'. I now check any word by doing a quick search for images. (Of course, that can cause problems too. There was the day that I wanted to know how to say 'sleeveless summer top' in French. I checked out the images and, an hour later, was congratulating myself on having bought only one.)

I wrote the vocabulary for each location—the garden, bedroom, kitchen—on plain cards and stored them in place, ready to check when I wanted to learn new vocabulary or look up things I'd forgotten. If the card became full, I would start a new one, leaving out the words I already knew well. I found this much easier than sticking labels all over the stove and window, as I remember my mother doing when I was at school. Labels don't tend to stick very well to soap and knives, rosebushes and underwear.

I translated recipes into French, with the instructions given in copious detail to maximise the vocabulary. I felt it necessary that the recipe remind me I would find the flour in the cupboard and that I needed to heat the butter on the stove and that I must wash the dishes afterwards. Every location in the kitchen made its way into at least one of the recipes I use regularly.

I started writing items on the shopping list in French if I knew the words. It slowly became entirely in French.

I constantly revisited my research on indigenous cultures and their memory methods to seek out ideas to help me with my French studies.

Songs reworded

Indigenous cultures sing their knowledge. Western cultures teach their children to sing the alphabet, and then cease the practice. Why do we grown-ups only go around singing 'I love you, I need you, I want you, I have lost you, will I ever love again?'—endlessly?

I could not get the tune of that pervasive French folk song 'Sur le Pont d'Avignon' out of my head, but rather than singing repeatedly about defunct bridges, I changed the words and sang about getting dressed. I sang that I was going to put on my skirt, and that I was putting on my skirt, and had put on my skirt. And I'm not bothered if Brother Jacques is still sleeping—he is welcome to miss morning bells for all I care. So I converted 'Frère Jacques' to be about washing body parts in the shower and drying afterwards. With constant repetition, at maximum volume, my vocabulary and use of tenses improved rapidly.

I made songs about the garden and the shops and anything else that had a long list of vocabulary. I enjoy singing loudly. It would be far better for the family if I were occasionally in tune. Entering certain rooms or doing particular activities triggered my brain to try thinking in French. Mangled French, I admit, but French.

I was doing at least an hour a day, often more. But I knew that I could be making serious linguistic mistakes and then re-inforcing them with every repeated conversation. I needed the reassurance of a formal French conversation class, so I started attending one every week and loved it.

In all these ways, I integrated French into everyday life and it soon became a habit.

Memory palaces everywhere

The house and the garden were effectively becoming memory palaces. I took the advice of Dominic O'Brien and set up what he calls 'gender zones' for all the vocabulary I would encounter outside the house.[2] When learning fruit and vegetables, I bought the feminine ones in the fruit shop and the masculine in the larger of our two supermarkets. For groceries, the masculine items were in the larger supermarket and the feminine in the smaller.

I would mentally chat to the appropriate bear when shopping, hopefully silently. There were a few occasions when I was so absorbed in our conversations about broccoli and brussels sprouts that I approached the counter with a cheerful '*Bonjour*'.

For the verbs I established a memory palace for each group. I put my regular '-er' verbs on a walk that took me down the street, along the tennis courts and around an oval.

I placed the verbs in alphabetical order. I found a list of the 100 most common verbs as my starting point. I extracted the regular –er verbs. *Aboyer* (to bark) was first, so I imagined a dog barking at me from the verandah of the house at the start of the walk. I included a small boy with the little dog, encouraging it to scare the old lady. This gave me the 'boy' to add to the letter 'a', which I know from the location in the walk. Adding emotion to the story makes it more memorable.

Some of the stories were quick to make up, some took a bit more effort. But I could always find some link to the French word.

By spreading the verbs over the kilometre or so of the walk, I left ample room to add more verbs at the appropriate location as my vocabulary grew.

If I want to describe weeding the garden, my brain instantly jumps to the appropriate location, which tells me that the verb starts with 'd' and *désherber* comes to mind. I know that somewhere in this memory palace I can dance and sing, ask, draw, listen, play, jump, work and study. I can commence things. But I can't finish them.

To finish is *finir*. Being an '-ir' verb, finishing requires me to mentally visit the '-ir' verb memory palace in the Botanic Gardens. There, I can do all the other regular '-ir' actions. I can act, choose, welcome, grow up, warm things, invade, cure, gain weight, lose weight and even grow old. If I want to do anything that is a regular '-re' verb, I need to go to the small park in town. There I can wait, defend, hear (but not listen), answer and even stretch. But I can't just be.

To be (or not to be), I have to go to the main shopping strip in Castlemaine, where all the irregular verbs hang out. There I can go, tell, want, make, know, and am able to do things. I physically acted out the verbs in each of these locations—only once for the action to stick and to know that I was conjugating them correctly—hoping that no one noticed.

Memorising vocabulary isn't enough

Much as I would love to claim that the memory techniques I've described are all you need to learn a foreign language, to do so would be ludicrous. Vocabulary is incredibly useful, but foreign language speakers don't put the words together in exactly the same way we do in English.

I cannot describe how much I enjoy French children's videos on YouTube. Just as I did as a child, the first few times through I sang the refrains easily and let a great deal of the other words float over me. Just like a child, I picked up common phrases and the way French sounds. Getting dressed in the morning turned into sing-along and dance-along fun. My first children's book was *On y danse les saisons. On y danse?* It apparently translates to 'Skip through the seasons'. Even French toddlers are already using expressions that do not translate literally.

I created as much of an immersion experience as I could, to learn those pesky in-between words and colloquial phrases. I soon advanced to adult songs in French. Edith Piaf's storytelling and clear pronunciation is a gift. I discovered that many of my DVDs had French dubbing and subtitles. *Fiddler on the Roof* in French is wonderful! I discovered the *News in Slow French* service, designed specifically for us learners. Technology has granted the chance to create an immersion experience, a poor substitute for the real thing but available to everyone.

Online courses

I tried some online lessons. Which is to say I *wanted* to do them. I *intended* to do them. I *planned* to do them. It's just that the actual sitting at a computer and allocating a specific time simply didn't happen. I know these courses work extremely well for other people. The few times I tried them they were very effective, but I struggled with being glued to an electronic device for another half an hour a day, given a writer's job is already hours in front of a screen.

Many people use Spaced Repetition System (SRS) software and techniques to memorise. They are really effective. Essentially, they force the repetition required for permanent recall, much as the repetitive teaching and the ceremonial cycle do for indigenous cultures. An SRS uses flashcards, either paper cards or on a computer. The image comes up with the foreign language word hidden on the back. These are repeated at regular intervals, depending on how well you know the word. If you get it wrong in training it will be repeated in the next session. If you keep getting it right, then it will be repeated at longer and longer intervals. You can add phrases to the flashcards, and sound if your flashcards are digital. If you are willing to commit time—such as while you commute to work on public transport—to a system which follows the same pattern every day, then SRS programs will be ideal.

My 'flashcard images' are all around me. As long as I chatter about what I am doing with my bears, then the necessary repetition is happening. I compile lists of new French words, which I check to ensure I'm adding new vocabulary. My methods probably won't be as efficient nor as rigorous as some others, like an SRS, but they seem much more fun to me. It will all depend on your personality and preferences and the urgency with which you need to learn.

Of course, you could just memorise an entire dictionary.

A very different language: Chinese

Malaysian memory champion Dr Yip Swee Chooi memorised the entire 56,000-word, 1774-page *Oxford Advanced Learner's English–Chinese Dictionary*. There are videos online of memory experts testing him. With a big grin, he answers every question, giving the page number for each word, either in Chinese or English, and the position of that word on the page. He then describes the way he has used over a thousand memory palaces to achieve this feat. In each palace, he has enough locations for the number of words on the page, typically between fifteen and twenty, but others with many more

Dr Yip mentored a Hong Kong memory athlete, Andy Fong, to become a Grand Master of Memory.[3] This honour is awarded to the very few people who can perform three specific feats: memorise 1000 random digits in an hour, memorise the order of ten shuffled decks of cards in an hour, and memorise the order of one deck of cards in under two minutes. You can see why Grand Masters are rare beasts.

I was privileged to meet Andy when he attended the Australian Memory Championships in 2017 as the international arbiter. Andy was less ambitious than his mentor: he memorised the *Oxford Elementary Learner's English–Chinese Dictionary* instead, which contains only 8699 words. It still seems like an impossible feat to me. I asked him why he would take on such a demanding challenge, and he replied:

There were a few reasons that I decided to memorise a dictionary. First, I wanted to better organise and familiarise my journey (or loci) to prepare for the competition that I was about to participate in. As I was aiming for the grandmaster title at that time, I need enough journey for at least 1000 digits of numbers and ten decks of cards. Although I had those journeys prepared, I felt the need to make it better organised so I can go through the journey as fast as I could. So, memorising a dictionary is one of the good practices.

Second, I wanted to memorise something other than cards and numbers. Memorising cards and numbers can be boring sometimes and I wanted to try something different. Third, it seems like a pretty cool demonstration. In particular, people link it directly to education and learning. This is something that is easier to explain memory skills on study than using cards/numbers as demonstration.

I stopped at page 215 after the competition in 2011. The first 215 pages have 2782 words (can't believe I actually counted them). It took me one day to memorise the first 50 pages. The 215 pages took about one month or so. But then I got bored, so I did the rest on and off. The memorising didn't take long but the revision could be lengthy.

The method I'm using is the journey method (or method of loci). Basically, each page has its own set of journeys and I put each word on each spot. So, when people ask me the fifth word on page 26, I would immediately recall the living room at my apartment where I was staying during my second year at university and see an 'apple' there, i.e., the fifth word on page 26 is 'apple'.

Andy then told me a reason for memorising that had never occurred to me.

We lost our baby after five months into pregnancy. It was a tough time for us as it was our first baby and we were looking forward to meeting him. After this bad personal experience, I've started memorising the Bible (starting with Philippians).

First of all, I am a Christian. I always wanted to use my skill to memorise the Bible, but I never had the motivation.

After the tragedy happened, my friend from church used a verse (Philippians 4:6–7) from Philippians to comfort me. Therefore, I started to read more Philippians and I found peace in a way, so I started memorising Philippians.

The thing with memory skills, if using them properly, is it can reduce the time of inputting a large amount of data. I can review the information anytime I want without reading it again, like on the train, day-dreaming or having lunch. It is the combination. Memorising the Bible helped me find peace and I can divert my focus to different things. It certainly helped me a lot during the difficult time.

Chinese and me

Enthusiasm blazing, I decided that I also needed a really foreign language to thoroughly test the memory methods. I chose Mandarin Chinese, as it is the language spoken by the most people. I started looking at Asian art, which I had always walked past in museums and galleries before. More fool me. I fell in love with the Chinese handscrolls at the New York Metropolitan Museum of Art, which I will talk, no, rave about in Chapter 6. I'll also rave about the Tibetan mandalas I was introduced to on the same trip. Chinese is the only language which can be traced from its preliterate past to the contemporary spoken language. How could I resist attempting to learn it? But I digress. This is the joy of trying to learn anything very different—those neural networks just get more and more complex and entwined.

I was starting from a position of absolute ignorance. Spoken Chinese was unintelligible to me and the script, although extremely pretty, was equally meaningless. Chinese has a variety of spoken dialects, including Cantonese and Mandarin. They are essentially different spoken languages sharing a written script. I found that baffling; which should I learn? But I have close family friends who speak Mandarin, as do the few Chinese speakers in Castlemaine, which confirmed my choice for me.

One major problem remained: where do you start with a language that has not a single hook that is familiar? I bought a number of textbooks and investigated a number of online courses. I came to the conclusion that your starting point depends on where you're heading. Some courses recommend learning to read pinyin, which is the official way of writing Mandarin in Roman script, giving the pronunciation; they say to start with speaking. Others recommend ignoring pronunciation initially and to start by learning the characters so you can read Chinese. Some recommend against attempting both at once, while others recommend diving into both straightaway.

Because I had a particular interest in the way the language and the script had developed, writing was clearly important. Because I wanted to engage with my Chinese friends and local native speakers, I also wanted to speak it. Because I wanted to use the memory methods to structure my approach, I just did it my own way.

The rule of thumb is that you need to know about 1000 characters for basic reading comprehension, and about 4000 to get well into the language. About 100,000 will give you all the characters in Chinese dictionaries, but I'd need a few lifetimes to get there. I knew none.

I was also starting from a basic misconception. I thought each symbol was a pictogram of a word. A little house would mean 'house' and a pretty mountain would mean 'mountain'.

But if that was the case, it would be almost impossible to represent all those in-between words that connect nouns and verbs to make a spoken language. It turns out that the vast majority of those gorgeous little square characters are combinations of signs, some of which give meaning and others of which offer the phonetic component.

I chose my hook, the radicals

When I found out the structure to the way characters are usually listed in Chinese dictionaries, I finally had a way in. The order is arranged by the radicals, the graphical building blocks of Chinese characters, which often give a clue to the meaning of the character and sometimes to the sound.

Officially, there are 214 radicals but my *Pocket Chinese–English Dictionary* had 187. Radicals are listed in order of the number of strokes used to write the character, so if I knew where a radical was in my memory palace, then I knew the number of strokes pretty reliably. I figured the 187 in my dictionary would be more than I could cope with anyway, so they became the foundation for my memory palace.

To begin, Damian and I went for a walk, and I chose a location for my first two radicals: man/person (*rén*) and woman (*nǚ*). I made stories linking the radical symbols to the houses and encoded the meanings. I couldn't wait to get home and look at some Chinese writing, and when I did I managed to find my two radicals. The totally unfathomable squiggles suddenly had recognisable fragments. I was extraordinarily excited.

I added more radicals to my palace, pausing at the relevant location, concentrating on the shapes on the page and in the building, fixing the shapes of the radicals really securely. After each addition to my memory palace, I sought out the radical in Chinese characters, which gave me a boost of confidence. The strange, square blocks of squiggles were giving up their secrets and I suddenly felt sure I would be able to decipher them all. Eventually. One day. Just not soon.

After a few weeks I had established the 187 locations in my Chinese radical memory palace, ready to link the radicals themselves. The locations are each a house block or so long, which enables me to add the words associated with that radical as I learn them. Each radical is a shape, which forms part of the Chinese characters associated with it. I can always find something resembling that shape on the house, shop, school or fence that provides the location for the radical. When I walk past a location that already has a radical attached, the significant window, door, roof or decorative verandah imbued with this shape always jumps out at me.

My Chinese radical memory palace is about five kilometres long, starting and ending near home. There are ample side streets, which can be brought into play if I need to learn the other radicals. I also have only used one side of the road, so I could extend to the other side if needed. For common radicals, like *rén* at location sixteen, I can attach hundreds of words to the house and garden.

I could go into endless detail about learning Chinese, but suffice it to say I have discovered children's videos online, bought children's books and textbooks and dictionaries and CDs of songs. I am working through a range of introductory textbooks, and continue to link radicals and words to their location in the memory palace, a few at a time as I come across them in my study. Among my favourites are the children's books on learning Chinese characters, because they teach how to do the gorgeous shapes simply enough for even me to follow.

But the most important thing to get me started was establishing a memory palace structured according to the way the information was presented.

A final realisation

I have many kilometres of memory palaces mapped out around town. They are helping to keep me fit, mentally and physically. You can also make miniature memory palaces in the form of handheld devices, which work extremely effectively, part of the extraordinary range of mnemonic art forms indigenous cultures all over the world have developed. But before we look at these in Chapter 5, it is time to learn another major lesson from our indigenous colleagues. We've seen that research shows that neuroscience and indigenous memory methods both advise the same things: your memory is going to be helped immensely if you use regular repetition, vivid stories, song, imagination, emotion and memory palaces. Indigenous cultures have mythological characters acting out every story, song and ceremony for a very good reason. Learning that lesson was some of the best fun of all.

CHAPTER 4

Characters, characters everywhere

You can try to just learn facts. You can read and reread dry columns of words or facts or dates. But you'll do far better at recalling those same facts if you have lively folk weaving stories around them. That sounds like you need to remember more than just the facts, but you'll find you do it so much faster and so much more effectively, that more to remember becomes less work.

Every practice I've talked about in previous chapters will be enhanced by creating really strong characters. The reason a visual alphabet is so much more effective than using memory aids, such as 1–sun, 2–shoe, 3–tree, is because the humans and animals are capable of *acting out* the stories, rather than just being a passive link.

I have an entire pantheon of characters who inhabit my

world through my imagination. My life is so much richer for their presence.

As every indigenous culture across the world will show you, vivid stories with lively characters are hugely memorable. I am really uncomfortable using the word 'characters' in this context, but we don't have an equivalent in Western culture. The usual English terms used when writing about indigenous stories are 'mythological beings', 'ancestors' and 'Old People', among others. The most appropriate words are those used by the cultures themselves. For the Pueblo, they are *kachina*.

I was entranced by the vivid and wildly varied Pueblo *kachina* from the moment I first learned of them. Spelled in a variety of ways in English, each *kachina* (or *katchina*, *katcina* or *katsina*) has its own specific design, which features on the masks and costumes of dancers at ceremonies. The designs are seen on pottery and petroglyphs and in the entrancing *kachina* figurines used to introduce them to children (see Plate 11). Although often called dolls, they are far more than playthings—they form the foundation of the Pueblo knowledge system. The more I read and heard and was shown at the Indian Pueblo Cultural Center in Albuquerque, the more enthralled I became with the entire concept and culture. I started collecting *kachina* made by contemporary Puebloled artists. A few are shown in Plate 12.

I was surprised to find that both Australian Aboriginal colleagues and Native American acquaintances tended to use the same expression when I asked them about these characters: they told me they are not gods but characters 'whose stories we tell'.

Māori ancestors

When the Māori sailed to New Zealand around 1280, they arrived at uninhabited islands. They successfully adapted to a totally different environment while maintaining the essence of their Polynesian origins. Their traditional knowledge system, *Mātauranga Māori*, involves a complex array of memory devices. The most significant is a carved wooden genealogy staff, known as *rākau whakapapa*. It has the ancestral figure at the top, with a series of knobs below each representing a generation. The ceremonial orator touches each knob as a memory aid for reciting the genealogy. At each knob, the orator may also recite the knowledge associated with each generation. In essence, the *whakapapa* organises the Māori encyclopaedia into a structure based on ancestry. The ancestors act as protagonists for the stories encoding an integrated teaching of all their practical, cultural and spiritual traditions.

The Māori require a large number of ancestors to have sufficient stories to store all the information. They typically recall about 25 generations of biological ancestors, over 600 years' worth of forebears. Beyond that, the human and mythological ancestries tend to be interwoven in the mythology. It is the knowledge that matters, not the specifics of every living person from a millennium ago. Initiates are expected to memorise the complex *whakapapa* accurately.

So, *whakapapa* works much like Australian Aboriginal songlines, except the structure is based in genealogy first and landscape second. All indigenous cultures store complex genealogies. The West African *griots*, for example, sing the history of the culture, including many generations of royal leaders. In these lengthy performances, the *griots* act as the cultural encyclopedia for their community.

Introducing rapscallions

Reflecting on the *kachina* and other indigenous ancestors made me think about the potential of using such characters in contemporary life. The ancient Greeks and Romans taught students to use characters to tell stories when memorising information. This is why the 2000-year-old stories of Zeus, Aphrodite, Athena and Arachne and their assortment of fellow actors are still so powerful today.

I wanted to create a set of my own ancestors with which to weave stories. But it would be culturally insensitive to use terms that may imply that I was adapting *kachina* or any other indigenous pantheon for my own purposes. At the time, I was working with artist and educator Paul Allen to explore ways in which we could implement ideas from *The Memory Code* in schools. I'll talk more about the implications for education in Chapter 8, but what matters here is our conversation about how to create characters with students without being offensive to the indigenous cultures teaching us so much. The solution was to coin our own term. We came up with lots of unusual names, checking for any existing association with a quick image search online—and repeatedly cringing at the vulgarity filling our screen. Our brilliant ideas were crossed out one by one with thick black lines.

Then Paul suggested 'rapscallions'. We loved the lively, cheeky connotations. Even more, we loved that an online search of it didn't lead us to any pornography. A few weeks later, we discovered the pure joy of listening to a five year old, with front teeth missing, saying 'wapscallion'.

So, rapscallions they are.

If you take only one piece of advice from this book, please let it be that you welcome rapscallions into your life. It doesn't matter whether they are based on family or friends or ancestors (biological or mythological) or pets (present or deceased) or toys (having resisted social pressure to put aside your beloved playthings) or historical characters or celebrities or complete figments of your imagination. Your life will be so much richer with rapscallions.

I have hundreds of rapscallions in action at the moment. You met two of them in the previous chapter, Fleur and Le Petit Professeur. There are another hundred who serve to represent numbers; you'll meet them below. And then there are my 'ancestors'.

My cultural ancestors

Singer Annie Lennox said, 'Our ancestors are totally essential to our every waking moment, although most of us don't even have the faintest idea about their lives, their trials, their hardships or challenges.'

How far back can you name your ancestors? How much do you know about them? I know who my parents were, and my grandparents and aunts and uncles. I know of a few relatives from earlier generations, but of them I know little more than names or where they lived or what they did. I have no knowledge of their lives and values, knowledge from which I can learn. By contrast, the ancestral characters who populate the stories of indigenous cultures are active participants in daily life.

With sincere apologies to all my biological ancestors, without whose private activities I would not exist, I have constructed a new set of ancestors whose stories can teach me.

I started by making a list of all the people from history I thought would serve me well. I enlisted my husband's help, knowing that his different educational background and better memory would add a richness to the selection. Over a number of weeks, Damian and I made a list of historical personalities. Damian knows a lot about economic theory, so he put in Adam Smith (1723–1790), John Stuart Mill (1806–1873) and John Maynard Keynes (1883–1946). These three, he said, and their historical contexts will serve as the foundation for examining the role of economics in history. And we would be able to link other economic ideas to them. That was the sort of logic we used.

We asked friends and family who should be included, which led to some incredibly interesting discussions. At one dinner I was asked why I had not included Averroës. Who? Never heard of him. I handed over my notebook, having no idea how to even spell his name. I now know he was a medieval Arab polymath whose full name was Abu al-Walid Muhammad ibn Ahmad ibn Rushd (1126–1198)—but I'll stick with Averroës. After a quick online search there was no doubt he would make the cut, no matter how many others had to go.

We ended up with a list of well over 300 names, which Damian and I attempted to whittle down to 52. This magical number would allow me to associate them with a deck of cards as a memory device, to ensure that I could put them in chronological order and not lose any. You could just use a memory palace and make it the right size for your final list, but I wanted to use cards, so 52 was the goal.

But we simply couldn't reject enough, because history has too many interesting people. We also wanted to include a range of professions to provide us with a hook for other historical figures in their fields. So I added a tarot deck of another 78 cards and we eliminated until we had 130 survivors, killing off many I would have loved to retain. There is one cheat: I have Buddha and Confucius on the same card, given they were contemporaries. And there are some horrible people on the list. You can learn from bad guys, sad as it is to admit.

We have scientists, artists, political leaders, conquerors and authors. There are no sportspeople nor film stars nor singers, although we do have composers. This is not some kind of value judgement; we just ended up with those we wanted to know more about. I expect that you would likely choose a very different set, but our list in Appendix D will give you a starting point.

I researched my ancestors. Staring at the cards gave me links. In the '3' on his card, I could see Buddha sitting cross-legged. In the '7' on Aristotle's card, I could see a leaning ladder from which he 'tottled'. Those images gave me the starting points for stories, weaving reality with fantasy. I have no trouble differentiating between the two. Sometimes I also use the pips on the cards and images on the picture cards for more details. Having a physical representation that I can hold gives me a stronger connection as I ponder their lives.

As I attached information to the card decks, they each became precious to me. Much more so than, say, Henry VI, who's part of my History Journey. Interesting as his story is—his ascension to the throne of England at the age of nine months and his tumultuous life, ending with his death at only 49 in the Tower of London—it just doesn't have the same impact as my ancestors. It is almost impossible to explain the difference emotionally between knowing *about* a historical person and learning *from* them.

Rarely a day goes by that one or more of my ancestors are not mentioned somehow. Before I gathered together my ancestors, I was completely unaware how often these people appeared in conversation or in the media, but now they have become part of my everyday life.

Ancestors in the History Journey

I put all my ancestors in the History Journey as well, which reinforced their order, complementing their link with the playing cards and tarot. At the birthdate for each, I could always find something to hook them onto—a water pipe for Aquinas, a column Occam slashes with his razor. Each of my ancestors took their place in history alongside their compatriots and in the midst of the events going on in the world at that time.

This raised all sorts of questions. How much did Michelangelo know about Copernicus's ideas? Or the others in the middle of the fifteenth century? The Ottoman Sultan Mehmed the Conqueror was doing his conquering. The Incan Emperor Pachacuti was building Machu Picchu and ruling a massive empire. Christopher Columbus was terrifying indigenous cultures and Leonardo da Vinci was doing his stuff. Joan of Arc was burning at the stake, and the apparently timid pacifist King Henry VI was on the throne of England while the Hundred Years' War raged on. Gutenberg's printing press was about to change the world.

There were a number of times when people asked me why I was staring intently at their property and scribbling on my list. At 300 CE, an elderly gentleman asked what I was doing. I tried to explain that his unit was the start of the Classic Maya period. He clearly concluded that I was a harmless nutter, smiled, nodded and rapidly retreated behind a closed door. At 1400 CE, a mother noticed me behind the bushes at the back of the school grounds. Not only did she accept my explanation but became quite enthusiastic about setting up her own History Journey.

A fifteen-year-old student, in a group exploring memory methods with me, wanted ancestors like mine. Reuben preferred to use a single deck of playing cards, so he reduced my list of 130 ancestors to his chosen 52 and prepared to set up his memory palace. We went to the local botanical gardens where I already had a French verb memory palace in place. I naïvely assumed that Reuben would simply use my locations. I thought that he would use the same key characteristics of the ancestors as I had and link to those locations. But none of that happened.

Reuben kept finding his own significant locations: a sign I had never noticed, a seat I had overlooked, the back of a building where I had used the front. His Cicero was orating, while mine was memorising. He was far more engaged and enthusiastic about creating his own memory palace, and his own interpretation of the ancestors, than he would have been simply learning mine.

This reminded me greatly of what indigenous colleagues had told me about oral tradition. The factual information encoded is robust and consistent. The methods used are robust and consistent. But the way an individual might elaborate and perform the story can vary greatly and skilled storytellers are hugely respected. The core information being conveyed never changes, however. Reuben's Socrates still died from poison, and his Shakespeare still lived in Elizabethan England, but the way he attached a story and encoded the information was uniquely his and much more memorable for that very reason.

The only way you will understand is to try it for yourself. The more I engaged with my ancestors, the more I understood how incredibly powerful the roles of the *kachina* and the Aboriginal 'Old People' and all the indigenous mythological characters were in every single day of the life of traditional people. We simply don't have this sort of personal engagement in our Western present with those of our pivotal past. We have lost a great deal.

The Dominic System for numbers

While my ancestors were in chronological order, and I could place them in about the right time in history, I wanted to know precisely when my ancestors had lived. Although I had placed them in my History Journey, that would have only given me their birth to within a decade or so. I wanted a method to remember the dates of their births, to know how long they had lived, and to know when they had died. I wanted to know who was influential in other fields at the same time, and what was going on in the world during their lives.

There are a number of methods for memorising numbers; the two best known are the Dominic System and the Major System. I tried both, but soon settled on the Dominic System, created by the British memory expert I mentioned in Chapter 2. (I'll explain the Major System in Chapter 10.) Dominic O'Brien says his system's title is an acronym: DOMINIC, Decipherment Of Mnemonically Interpreted Numbers Into Characters. Most of us in the memory world just think of it as his name.

The trick is to assign characters to the numbers 00 to 99. That way, abstract numbers become characters who can act out stories. When Dominic first created this method, he assigned characters to any numbers that were already significant for him. He linked 10 to the British prime minister, who lives at 10 Downing Street. Another number was the year he was born. Once significant dates had been assigned, he listed the leftovers. For these, he assigned letters to the numbers and associated them with the initials of friends or celebrities. The number 11, for example, became AA for Andre Agassi. Dominic's code was:

1=A	6=S
2=B	7=G
3=C	8=H
4=D	9=N
5=E	0=O

When I set up my own version of the Dominic System, I found very few numbers suggested anything to me instantly and my preference for rigid systems led me to assign characters for all the numbers according to letter counterparts. I chose letters that suggested to me the shape of the numbers:

1=J	6=G
2=R	7=T
3=B	8=L
4=A	9=D
5=S	0=O

Why, I hear you ask, did 8 end up as L? Good question. I can't remember. I imagine it as a shovel loading lumps of stuff, but that's not why I started with 8 and L. You should use the letters that the numbers seem to suggest to you.

For 'DO' I immediately thought, unsurprisingly, of Dominic O'Brien. 'JA' became Jane Austen and 'RR' became cartoon character Roger Rabbit. The definition of celebrities and friends can be as broad as you like. Even so, I couldn't manage good character associations for every combination, so I had to fudge a few of them. For 'OO', I ended up using Old Oldfield. I'm a huge fan of Mike Oldfield's *Tubular Bells,* an instrumental album that he recorded so long ago that he must be old by now. It works for me.

You must decide on your own characters for the 100 numbers. Anyone else's friends and relatives will be meaning-less for you, even if you recognise the famous names. Their list will have far less emotional impact and will therefore be less memorable.

Put in plenty of horrible people. Horrible people give you much more memorable images when combined in the way described below. Sad, isn't it? Some of my horrible people are politicians or historical tyrants. Some are those I dislike in everyday life: people I meet socially or, dare I admit it, family members. To their face, I am perfectly polite and agreeable but in my imagination I have total freedom and privacy. Sometimes I am dismayed by the gruesome images in which they become embroiled, but it can be immensely satisfying.

For years of birth and death, you have four numbers. To memorise them you need to use a Person–Object method. (In Chapter 10, I'll describe the way I use the Dominic

System in memory competitions in groups of six numbers using a Person–Action–Object method.) For Dominic O'Brien, for example, I would have assigned a deck of cards, because my most immediate image of him is memorising card decks. That proved a problem for me because I already had a magician friend dealing out cards for 81. Every one of the 100 objects must be different, or a deck of cards could mean 81 or 90. So I went to the internet and found the first image of Dominic on offer. It happened to be of him holding a picture of a plastic bag with what looks like carrot chips in it. I have no idea of the story behind the image nor what is really in that bag, but in my imagination, Dominic spends his life constantly filling plastic bags with carrot chips.

To add the date to any event, you simply use the two characters together. Scientist Rosalind Franklin was born in 1920. I want to commit that year to memory. Franklin has the Stars card in the arcana of the tarot deck. That card has a picture of a lady throwing stars so, for me, that lady becomes Rosalind Franklin.

The number pair 19 (in my system, 1=J and 9=D) is the initials of a very dear friend and mezzo soprano, Jacqueline Dark, a star of Australian opera. I need my image of Jacqui to be doing something to get a vivid active image. That was easy. Jacqui sang the role of Mother Superior for a year in *The Sound of Music*, so my image of her is in a nun's costume. Jacqui is well known for her bawdy operatic cabaret, so this image is very funny and memorable for me. It would be of no use to you.

Number 20 is Roy Orbison (2=R, 0=O), whom I imagine in his trademark dark glasses.

Franklin's birth year, 1920, requires me to link Jacqui, the person, to an object, Roy Orbison's dark glasses. So Jacqui became a nun magnificently singing 'Climb Every Mountain' while wearing dark glasses. Rosalind Franklin is in the audience, tossing her stars at the stage in appreciation of Jacqui's performance. I can see it so clearly!

I then do the same for Franklin's death in 1958. The number 19 remains the same, my friend Jacqui the singing nun, and the number pair 58 is SL (5=S, 8=L) who I decided was Sonny Liston. So now Jacqui needs to wear boxing gloves.

To memorise the years of Rosalind Franklin's birth and death, I imagine Jacqui wearing her nun's costume and The Big O's sunglasses, singing about climbing every mountain. Next thing we know, Jacqui is heckled by an audience member. Without missing a beat, she dons boxing gloves, knocks them out and starts singing about crossing every stream.

That is one way to do dates and it will work well. In fact, as I have put everyone in the History Journey, which gives me their century, I don't need Jacqui at all to remember Rosalind Franklin's dates: I know that Franklin was born in the 1900s from her location in the History Journey at the back wall of our garden. I only need to remember 20, from her birth in 1920, and 58, from her death in 1958. So I have Rosalind Franklin tossing stars while Roy Orbison serenades her. She then confronts the other DNA scientists who took all the glory for discovering the double helix structure, donning boxing gloves to exact her revenge. From 1920 to 1958 is only 38 years. Why did she die so young? Every time I play with the information, I want to know the answers to more questions.

Friedrich Nietzsche was born in 1844 and died in 1900—from 44 to 00—but the change in century is no problem: if the latter date is a lower number than the former, it must be in the next century. I have Friedrich Nietzsche philosophising from the stone fence of a house across the road from home, so I already know he was born sometime between 1840 and 1850. To remember his year of birth and year of death, I need to add together my person for the number 44 (4=A, 4=A, Andre Agassi) and my object for 00 (0=O, 0=O, Old Oldfield). It seems that Andre Agassi isn't terribly impressed with Nietzsche's philosophies and serves up a tubular bell full speed to shut him up. I can now retrieve those years any time I want them.

I make these associations at a rate of perhaps a few a day, and revise them later that day, the next day, in a week, and again in a month if I need to. The images flash into my head momentarily when I walk the History Journey and it is exactly that, just a flash. That's the beauty of the system: I don't decode the image unless I need it. I have an endless array of images which I can call up and decode at will. As long as the image exists, I can always work out the dates.

While I was working on my memory systems, Dominic sent me small pieces of advice. I refer to them as my Dominic Dripfeeds. He wrote: 'It is not necessary for you to conjure up a perfect photographic image of these people. You just need to recognise them for what they represent. The best way to do this is to assign an action and prop to each person.'

This is really useful advice, because many of my ancestors look similar. Most of my Dominic System people are sort of blurry, with a single object dominating the image. Anything active is more memorable than a static image, so Old Oldfield is a fuzzy Mike Oldfield vigorously striking that iconic tubular triangle. The characters become caricatures: Donald Trump's hair is even more wildly exaggerated while Annie Oakley shoots anything within range. I can always add an accurate image of their appearance later, if needs be.

Characters in the stars

There isn't a culture in the world that hasn't populated the night sky with a pantheon of characters. Whether it is the stories that make the arrangements of the stars memorable, or the arrangement of the stars that make the stories memorable, I really don't know. I suspect it is a bit of both, a sort of feedback loop. There are thousands of examples of deep knowledge of the skyscape, from traditional cultures all around the world. Patterns found in the stars, and in the dark spaces between them, have been used to represent characters for tens of thousands of years and probably much longer.

The Ancient Greeks used the position of the stars as a memory palace. One of the most famous Greek memory experts was politician Metrodorus of Scepsis (c. 145–70 BCE). His memory so impressed Cicero, Quintilian and Pliny the Elder that they frequently mentioned his extraordinary abilities in their writings. One technique used by Metrodorus was based on the zodiac, which was at that time divided into twelve signs, then further divided into 36 decans, each represented by a character. The decans were divided into ten degrees. It is thought that Metrodorus imagined ten different backgrounds, which he applied in each decan. The character moved through these ten locations in order. Metrodorus therefore had 360 locations to use for memorisation, each unique, with its own character and background. His memory palace was permanently kept in order by the positions of the stars.

Just like the landscape songlines I talked about in previous chapters, Australian Aboriginal cultures use the skyscape to navigate. They don't simply follow a memorised star chart, they also use the star patterns as a memory aid through story. The skyscape narratives may mimic those of the landscape, the two working to reinforce each other.

Across the globe, cultures have used the star cluster known as the Pleiades for stories. In Australia, they feature as the final representation of the Seven Sisters songline: the traditional owners describe the story of a predatory, lustful and previously rejected loner pursuing the seven Ancestral Women. He initially took the guise of a man, who to this day chases the seven sisters over land and sky, eternally linking the two. As explained by *Kungkarangkalpa: Seven Sisters Songline* project:

An Australian songline is a Aboriginal voice map of country. The song follows the footsteps of the ancestors of the Tjukurpa Dreaming. Wherever their feet trod they created waterholes, hills, trees, bush foods, wind and rain.

The country is alive with songlines criss-crossing the vast deserts and green coastlands of this country. It has been sung into being in the beginning and continues to be sung today.

Songlines are cultural webs of memory, intricate maps in the mind of country. The songlines, like web-links, carry spiritual, ecological, economic and cultural knowledge between peoples across Australia.[1]

The Seven Sisters songline tracks through country of many Aboriginal cultures including the Martu, Anangu, Pitjantjatjara, Yankunytjatjara and Warlpiri. It covers over 500 square kilometres of this huge continent. This must be the largest memory palace in the world. It's not only a memory palace through country but also across the sky, as was wonderfully demonstrated in a substantial museum installation at the National Museum of Australia opened in 2017. Viewers lay on couches under a domed screen. The audio track enabled us to hear the Elders tell the story as we watched the landscape and skyscape change around us.

The stories are told and sung and danced. They are bawdy, emotional, frightening and humorous, stacked with adventure and tension. They are everything a good narrative should be.

The skyscape could work as a wonderful memory aid, just like the landscape. I have tried using the skyscape for this purpose, but found it inconvenient and complex compared with the ready availability of the landscape. At this stage, I am leaving the mnemonic beauties of the skyscape to my Indigenous colleagues, who are much more skilled in that arena.

Back in 2017, as we left the darkened room of the Seven Sisters songline dome, we were dazzled by the colours and sounds of traditional art representing the seven sisters and their story in paintings and weavings, on spears and in sculptures. So many *objects* also act as integrated memory devices for the complex story.

Portable memory devices are used by indigenous cultures around the world. Some of the best fun I have had has been adapting the lessons from the way traditional objects are used so that I can apply them in my contemporary life.

CHAPTER 5

Weird and wonderful portable memory aids

The landscape is a wonderful memory aid, but it's just the beginning. Using that incredible imagination of yours, almost any physical object can become a memory aid as long as it has a sequence and variety.

Indigenous cultures use a huge range of portable art to make their knowledge memorable. In this chapter, I will present the methods I have found most effective, from belts, bracelets and decorated boards, to notched sticks, bundles of objects, food dishes, knotted cords and even your own body. Take your pick!

The Incan *khipu*—memory devices composed of garlands of knotted cords—were so effective the Inca could rule an empire without a written script. The Inuit of Greenland carved wood into tactile maps that acted as portable representations of landscape memory palaces. The extraordinary navigators of the Pacific Islands crossed thousands of miles of open ocean without any charts or navigation instruments, depending on knowledge encoded to a stick chart consisting of pieces of wood tied together, sometimes decorated with shells.

The Australian Aboriginal *tjuringa* (or *churinga*) is a decorated wood or stone object only used in restricted ceremonies. The patterns are sacred and owned by those initiated into the knowledge of producing them and what they represent. By enforcing such restrictive rules, the critical information has been maintained accurately over the years, decades, centuries and millennia. Without restricting the training and recitation of knowledge, the content could be corrupted by the 'Chinese whispers' effect.

The Australian Yolngu clans of north-eastern Arnhem Land describe their figurative and abstract designs as representing the pathways taken by the ancestral figures, who created the land and sky and everything within them. The ancestors named every animal and every plant, every part of the landscape and every star, providing the language for everything that might be encountered and all the associated knowledge. The artworks are just one component of a complex knowledge system called the *madayin*. The suite of memory devices used by the Yolngu integrates songs, dances, stories, ceremonies, the landscape, skyscape, paintings and many decorated objects.

Absolutely enthralling are the birch bark scrolls (*wiigwaasa-bakoon*) of the Ojibwa (Anishinaabe) people of North America. Adorned with geometrical patterns and shapes, the scrolls are gradually unfurled during a performance. Ancient Chinese, Japanese and Korean narrative scrolls were also unfurled slowly as they were read, as I'll describe in the next chapter. The beautiful objects acted as memory devices for vast stores of knowledge.

By copying these mnemonic technologies, I am beginning to glimpse what indigenous cultures mean when they talk about the spiritual power of their sacred objects. I was not prepared for the emotional impact I would feel when holding my memory boards. I feel this most strongly with my first memory board, adorned with its beads and shells. It was inspired by the Luba people of the Democratic Republic of the Congo and their incredible *lukasa*.

The *lukasa* of the Luba people

The *Luba Bumbudye*, which translates as 'man of memory', were members of the powerful, elite and secretive society called the *Mbudye*. These men were essentially the tribal encyclopedia. Researchers have recorded that they could recite the history of their nation; genealogies; lists of kings; migration stories; royal political practices and etiquette; techniques for hunting, smelting, blacksmithing and other critical technologies; knowledge of the movement of the sun, moon and certain stars; cultural and social protocols; behavioural expectations; lists of deities and ancestral spirits; and the sequence of their complex ceremonies—and it was all encoded to a small decorated piece of wood just like those in Plate 13.

Even knowing that they trained for years, those claims seemed to me to be, frankly, wildly exaggerated.

So I grabbed a bit of wood small enough to be held comfortably in one hand (also in Plate 13) and I glued on some pretty red and green beads and nice cowrie shells. I didn't even think about what I was going to encode, I just made it look pretty. The technology wouldn't work anyway, so why waste too much time on it?

I was astounded.

Even with such an embarrassingly sloppy approach, my first *lukasa* worked an absolute treat. (I did even better with my second *lukasa*, which I designed to suit its data. It is also shown in Plate 13 and will be explained in the next chapter.) Each bead needs to be distinctive, so I strongly recommend using a variety of shapes and colours, but there are ways to distinguish them even if you are using identical beads. For example, the grain of the wood around where you place each bead can be quite distinctive if you look closely. The beads are also differentiated by those around them and the gaps between. Each bead will have a different distance to the edge of the board, so when you hold the board and touch the beads the position of your hands will change. All of that becomes valuable when adding meaning to apparently mundane objects.

The process of encoding knowledge to beads and shells and marks on a bit of wood will be very similar, no matter what topic you choose for your memory board. As soon as you have your information structured into some sort of order, you are ready to go.

As an example, I'll describe the way I memorised a field guide to the birds of my home state of Victoria encoded to my first *lukasa*. To make the board, I used some beads and shells from old necklaces, so it cost me almost nothing. Its value is immeasurable.

Encoding the birds

I am married to a very keen birdwatcher, my motivation for choosing this topic. Out birding with Damian one day, he pointed to a small bird with yellow on its wings and a black-and-white front. New Holland honeyeater, he told me. I saw another. Or did I? How could I be sure that it was another New Holland honeyeater and not some other bird that looked very similar, perhaps one I didn't even know existed? I won't elaborate on Damian's frustration in trying to teach me to identify birds.

There was only one solution. I needed a complete list in my head.

There are 412 birds in 82 families in the taxonomic list deemed acceptable to most serious birders. I started at the first bead on the top left-hand corner and associated it with the first family on the list, the Dromaiidae. This family has only one bird in it, the emu. All the families end in -idae so I only needed to get the Dromai- bit associated with the bead. As it was the first bead on the *lukasa*, I imagined a drum roll. I imagined the emu pecking at the drum was creating my drum roll. I just had to adjust the pronunciation of 'Dromai' a little bit. A few weeks later, that bead fell off the board, leaving only a sharp piece of glue behind. I was about to stick the bead back on when I realised that sharp point would remind me of the emu's sharp beak and aggressive behaviour. The emu is now permanently represented by my failure at kindergarten-level gluing.

I had stuck about 120 beads and shells on the board so I mentally grouped some of the smaller beads. I had to associate the tiny pipit with one of these groups of tiny beads. The pipit family is Motacillidae, and the group of beads, if you look at them sideways and squint a bit, looks sort of like a motorcar. My work was done.

Halfway through the families, near the edge of the board, is the extraordinary Australian lyrebird, family Menuridae. It so happened that when I attached the bead some glue dribbled down the side of the board. There are advantages to being pitiful at crafts, because my imagination immediately associated that dribble with *men uri*nating. My brain refused to forget that association, despite my best efforts to use something more mature.

Making up the stories is a key part of memorising, but they need to be your own. The Charadriidae family—plovers, dotterels and lapwings—is linked to a story about my great-niece Kara, because the taxonomic name is pronounced 'Kara-dree-idee'. The story is about my great-niece's guitar playing—far more meaningful to me than a story about some celebrity or historical character and of absolutely no use to anyone outside my family. Good stories will come to you very quickly once you are used to letting your imagination make connections.

Once I had all 82 families associated with the beads, I started adding the individual species within the family.

The first large intact bead in the top left-hand corner is green and yellow, and has a little gold speck on it. The second family to be encoded was the Anatidae: ducks, geese and our wonderful black swan. I decided that speck of gold looked like a little duck tail wagging. Okay, that took a bit of imagination but now I cannot look at that bead without seeing that wagging tail.

Sixteen members of the duck family needed to be attached to that one bead, which took a story. The first duck is the magpie goose and a few further down the list we have the swan. Two Australian football teams are Collingwood (the Magpies) and the Sydney Swans, so the story became a tale of a football match that ends in a massive brawl. The names of the first eight ducks became the players involved in this fight. Two *tea* ladies came onto the field at half-time (the grey *tea*l and the chestnut *tea*l) and the Australasian shoveler dug graves to bury the dead. The musk duck wore musky cologne for seduction in the stands, while the hardhead clobbered the pink-eared duck around the ears and left others black and blue with bruises, the blue-winged and the blue-billed ducks.

I chose to remember the sixteen ducks in that order because those close to each other in the narrative are also similar scientifically. Those in the same genus I turned into partners, like the two tea ladies. Those tags just give me a bit more information. I now have sixteen duck-characters on which to build more and more information about their identification and habits. And to amuse myself with their story.

I didn't add the species to the families in order. I added whatever family I saw most often. So the honeyeaters on bead 51 were encoded first, the albatrosses and shearwaters last. When I added one species into a family, I would add them all.

One day, Damian saw a bird he couldn't identify—not a very common occurrence—but he knew it was a honeyeater. I was able to reel off the 36 Victorian species, and by process of elimination, he was able to identify and then confirm that we were looking at a juvenile friarbird.

As I worked my way through the 82 families and 412 birds, I kept expecting to strike trouble making up images or puns or other ways to make the families and species memorable. But whenever I pondered for a moment, playing with the name and letting my mind wander, I always found a link to the bead or shell, or the position on the board, or the grain of the wood or some glue that had gone astray. I did print a photo of the *lukasa* to annotate when I started but I soon found that it wasn't necessary. I could always refer back to the original taxonomic list. The connections between birds and beads stuck remarkably well. It became a delightful game.

It's not just the look of a memory board that will serve your purpose. As I recite the board, I naturally touch each shell or bead. Their feel and position on the wood helps to make each unique. An illustrated design will still work, but you can make a *lukasa* even more powerful with a tactile response.

As I touched each bead and said the family name, a rhythm emerged. Dromaiidae, Anatidae, Megapodiidae, Phasianidae, Podicipedidae . . . The rhythmic chant became a song. I often sing my bird song loudly in the shower. It sounds fantastic to me, although family members have suggested otherwise.

About a quarter of the way into the song I get to the cockatoos, family Cacatuidae. There I imitate the raucous call of the cockatoo, 'Caca-caca-caca-tuuuuu-idee!' One of my granddaughters, Leah, about eight years old at the time, sang the rhythm in mumbles with a clear 'idee' on the end. She finished her shorter version by belting out a superb imitation of my attempts at a cockatoo. Leah had no idea what all the words meant, she just liked the rhythm and making a lot of noise.

I've been told by senior members of different cultures that restricted songs might be known to uninitiated members of the community, but the true meaning would not be revealed until initiation. One method of keeping information restricted is to use archaic language, which is only taught to those who have rights to the information. This reflects what happened with Leah: the archaic language of scientific taxonomy is unfamiliar to her, but that didn't worry her in the slightest. She will learn the meaning later should she wish to become initiated into the birding fraternity.

I started to get more 'indigenous' in my encoding. The sanderling is a tiny bird that feeds in groups at the water's edge, running in and out with the waves, collecting insects from the wet sand. When I sing the sanderling, I dance up and down imitating their dance. I loved doing that little jiggle so much that I started other dances for birds with particularly distinctive movements. For those whose call was significant, the song became more elaborate still. I laugh for the laughing kookaburra and warble for the reed warbler.

Why sit still in silence to study? Why not make up stories and sing and dance and make a lot of noise? Maybe not in the library, but certainly in the shower.

My *lukasa* soon became so familiar to me that I didn't need to have it with me in order to use it. I occasionally sketched it to burn it into memory, but I now have it with me all the time, safe in my imagination.

Adapting for change

There's always the possibility that the information you have linked so wonderfully to your *lukasa* will be changed. This is not a problem. You just need to add to the story. Do not try to un-memorise (also known as forgetting) what you'd previously learned—it won't work. You need to incorporate the change within the story until eventually the old knowledge fades and you are left with the new.

Which is handy, because taxonomists like to keep changing things. In their great wisdom, they decided to absorb the family Cracticidae into Artamidae. The latter are small, shy wood-swallows who are nothing like the much larger Cracticids, our confident songsters including the magpie, butcherbird and currawong. So I added to the story—it became a battle, a David and Goliath epic, in which the small, delicately coloured Artamid army defeated the noisy, big Cracticids and absorbed them into the Artamid colony.

I still think they should be separate families.

Adding new information is just a matter of adding more to your stories. If there is so much information that your story has become too complex, then you can link it to a memory palace. I have done that for our honeyeaters. The 36 birds encoded in a single story to a single bead was fine until I wanted to add lots of detail about identification, behaviour, habitat and breeding for each species, many of which are very similar. So I created a honeyeater memory palace. The board and the palace work together seamlessly.

Or you could use multiple boards.

It was one of those 'Why didn't I think of that?' moments. One of the readers of *The Memory Code* took the concept of a *lukasa* and danced in her own direction. Swedish resident Julia Adzuki was on a visit home to her native Australia. I was enthralled as she revealed her unique approach. Julia has developed a set of fifteen boards that fit magically inside a wooden box. She uses these *lukasas* when teaching the fifteen different classes for the Skinner Releasing Technique. The classes involve the teacher verbally describing images to students who then respond in movement. Julia might describe a poppy unfurling when considering the curling and uncurling of a whole body in dance. Each of the movement-prompting images needs to be offered to the class in the correct sequence.

FIGURE 5.1 Julia Adzuki with her stack of memory boards. (LYNNE KELLY)

Julia had previously needed to memorise ten typed pages of text for each class. She made a *lukasa* for each of the classes, each adorned with around twenty items. Julia had collected many of the beads, small stones and shells in Turkey during her own training, which made them more significant and memorable. My favourite was the broken shell with the coil of the centre spine rendered visible. She used that half shell as a memory aid for the part of a class relating directly to the human spine.

Memory boards galore

You really can make memory boards in an amazing range of formats, and using lots of different materials, not just imitating the *lukasa* with beads and shells. Most of my boards are wood, but either carved, burned with a pyrography pen, painted or decorated with ink. I like experimenting. The most important thing is to not hurry them. Taking your time will be faster in the long run because the more you ponder a symbol, the more firmly it fixes in memory.

The majority of my memory boards are flat pieces of wood. A few are spheres. Other people are more imaginative—readers have emailed to tell me they've used their mobile phone and tablet covers. Much as I would love to wax lyrical about every one of the boards I have made over the last few years, you would find it rather repetitive, so I'll mention just a few.

I love spiders. I am a recovered arachnophobe who overdid the cure. Creating my spider memory board was a very different process. For my bird *lukasa*, I wanted to know every bird species in Victoria, the official 412. There are just a few more spiders. Around the world, there are over 40,000 named spider species. In Australia alone, there are about 3500 named species, with many times that not yet classified. Most of these spiders I will never see and certainly never identify. They are tiny and need an expert staring down a microscope focused on their genitalia to recognise their species.

I'd be happy to just be able to name a garden orb weaver or a jumping spider, and perhaps get more specific for the common species—a knobbed orb weaver or a white-moustached jumping spider. But mostly, I just wanted to be able to identify a spider to family level.

There are 117 families in the world at the time of writing. The critical identifying feature for spider families is the eye pattern. So I drew 117 tiny little eye patterns in ink on my memory board, in the order they are given in the taxonomic references (see Figure 5.2). I added some fancy swirls for every

FIGURE 5.2 Top: Spheres for ceremonial cycles, carved by Tom Chippindall and modelled on the Neolithic carved stone balls from Scotland. Bottom: One side of my memory board for spiders. (LYNNE KELLY)

fifth species, which made the board more memorable. For the families I encounter most—jumping spiders, orb weavers, wolf spiders and my various house guests—I have space to add swirls and curls and spikes and squiggles for particular species. I now know the eye patterns for the major families and am constantly adding to the data. I have a hook for every spider I read about because arachnologists always talk in terms of the families and the names are now all familiar.

Ceremonial cycle balls

All the indigenous cultures I have explored have a ceremonial cycle; their songs are repeated regularly to ensure none are lost. Repetition is essential for information to be retained reliably in long-term memory.

I have created a lot of songs. Although my Countries Journey tells me the countries in order of their populations, I confess that for many of them, I had no idea *where* they were. Although I could have encoded the data to the memory palace, I decided to create songs to help. I sing the countries in Africa, starting in the north-west corner and going clockwise right round the continent with the odd dash inland, and then back around the perimeter for the islands. I have another song for the islands of the Caribbean, another for Micronesia and so on. I have my song for my bird *lukasa* and a whole swag of songs for French and Chinese vocabulary. I don't want to forget any of them.

Friend and luthier Tom Chippindall carved wooden spheres with bumps and etchings (see Figure 5.2). These were modelled on the carved stone balls of Neolithic Scotland, one of which is shown on the cover of this book. I was testing whether those objects would function as memory devices for my PhD research.

To encode my song cycles, I selected a ball from the set and decided which songs I would assign to it. As I found with the Neolithic carved stone balls when I examined them in two Scottish museums, there is always a simple way to follow a sequence in the decorations. The middle ball in Figure 5.2 was easy just because of the different etchings and simple layout. I use it for some of the longer songs, such as the bird *lukasa*, each song assigned to a knob or crevice.

The balls at top left and right of Figure 5.2 have multiple knobs, which look identical at first glance. When you hold them (the originals or the copies), a pattern in the layout emerges that enables you to hold it with the same orientation every time. By choosing a specific knob as the starting point, the others can be followed in sequence due to the pattern. It may be that the original Neolithic balls were painted to make each knob different. I haven't found this necessary. With use, I found that each knob is uniquely identifiable. This might be due to its position in my hand as I hold the ball or the slightly different feel as I touch it. I use the ball on the left for short songs from many of my memory experiments. About half of the ball on the right is for my language songs. The other half is for world geography, where I use a knob for each verse. For example, my Africa song has four verses that are encoded to four sequential knobs.

I work around each carved ball over a week or so, singing each song in the sequence defined by the wooden balls. I sing in the shower, when cooking or gardening, or while walking in the bush. Every few months at least, each song will be sung. It is my ceremonial cycle.

The carved balls work wonderfully well. Almost any shape can be turned into a memory device.

Genealogies in wood

Indigenous cultures often record their complex genealogies in a variety of memory structures. I have been able to replicate, at some level at least, the way memory systems work for all the other genres of pragmatic information. But I can't see how to memorise even the simplified network diagrams of Australian Aboriginal relationships.

I had dinner one evening with Dr Tyson Yunkaporta, who is a Bama man. Yunkaporta told me that Aboriginal genealogies resemble a relational database. Everyone is connected by their relationship to others, and to the landscape, so the only way you can really see the genealogy is from within it. The genealogies record peoples' relationships through blood ties but also by the link to totems and clans and marriage and all sorts of associated responsibilities. I tried to think of my own family that way but it just got too complex, so I gave up.

In my own experiments, I have resorted to a simple linear tree going back generation by generation such as those used for royal or chief lines. Relationships within the entire tribal groups are recorded differently. I was able to replicate, at a simplified level, the way the genealogy staves of the Māori and Iroquois work.

A beautifully carved stick, or even one that you have just roughly hacked, will work as a memory device. Indigenous cultures all over the world have notched or painted sticks to be used as message or calendar sticks. Many of the aesthetically beautiful and culturally powerful carved canes of Native American tribes act as complex mnemonic devices. Some record the results of war: those killed and taken captive. Others record history, treaties and mythology. Some of these 'sticks' are so large that you are more likely to refer to them as posts, but many were portable.

The Māori wooden genealogical staves known as *rākau whakapapa* are still in use. They consist of a sequence of carved knobs, each representing a generation. *Rākau whakapapa* are usually just over a metre in length. The orator touches each of the generations as he recites the *whakapapa*. His performance is respected for both the artistic achievement and the exceptional feat of memory.

The recitation is not just simply reeling off the names. For each of the forebears there is a great deal of associated information, including origin stories, which describe the creation of all aspects of the universe and all living creatures. Māori can trace their ancestry back more than 800 years—that is, over 30 generations—to the first settlers who arrived by canoe from Polynesia.

All the names of the Iroquois Confederacy are recalled through the use of a decorated stave. The Confederacy now serves the Mohawk, Oneida, Onondaga, Cayuga, Seneca and Tuscarora Nations. The mnemonic device is thoroughly documented in an intriguing booklet, *The Roll Call of the Iroquois Chiefs: A study of a mnemonic cane from the Six Nations Reserve*. Two sections of the annotated diagram of the cane are shown in Figure 5.3.

Written in 1950, William N. Fenton's research with Cayuga colleagues revealed a mass of data about the cane and its last owner, Andrew Spragg, who had been born a century earlier. Spragg was a famous Cayuga ritual singer who would

FIGURE 5.3 (OPPOSITE) Details of the mnemonic cane from *The Roll Call of the Iroquois Chiefs* by William N. Fenton. (SMITHSONIAN INSTITUTION)

FOUR BROTHERS SIDE
OFFSPRING (NEPHEWS)

THREE BROTHERS SIDE
SIRES (UNCLES)

MOHAWK

Newhouse 1885 M.   Charles 1917 Oa.

Dekarihokenh	1 Dega'iho'gen'
Ayonhwathah	2 Hayen'wen'tha'
Shadekariwadeh	3 Sha'dega'ihwa'de'
Sharenhowaneh	4 Shaenho'na'
Deyoenhegwenh	5 Deyon'he'gwi'
Orenregowah	6 Oenhe'go'na'
Dehennakarineh	7 Dehenna'ga'i'na'
Rastawenseronthah	8 Ha'staven'sen'tha'
Shoskoarowaneh	9 Shoagoha'ihan

ONEIDA

Odatshedah 10	Ho'datche''de'
Kanongwaniyah 11	Ganon'gwen'yo'don'
Dayohagwendeh 12	Deyo'ha'gwen'de'
Shononses 13	Shonon'ses
Dehonareken 14	De'na'egen'/a'
Adyadonneatha 15	Hadya'donnen'tha'
Adahondeayenh 16	Dewada'hon'den'yonk
Ronyadashayouh 17	Ganiya'dasha'yen'
Ronwatshadonhonh 18	Honwatca'don'hwi'

1 *a* Hai Hai	*b* Five Nations denominated by Leading Chiefs (preface to Buck version)
b Take up path	
c Excuse errors of sequence, omission as anciently performed when all words were together	*a* Completed League?
2	
b Heads in graves	
3 Beneath your head	*a* You have taken it with you into your graves
What you decreed	

take the cane with him when moving between tribal locations to sing the history. Spragg would press his thumb against each peg to call out the name of a particular chief. There were originally around 50 pegs on the 90-centimetre cane. But this is no simple list of tribal leaders. Pictographs associated with the pegs represent plants and birds, anthropomorphic designs and human body parts along with abstract symbols.

One side of the cane gives the list of chiefs, two sides of the cane relate to relationships with the other members of the Nation, while the back of the cane has symbols that refer to history, ethical expectations and the law. This is, of course, only the public knowledge. Like all traditional mnemonic devices, there will have been much more complexity for those initiated into its use.

It seems that it took two years of bribery before Spragg finally gave up the cane. Cayuga chiefs have since expressed their disappointment that Cayuga Nation traditional property has left the reserve and the original notes that came with the cane have been lost. Without the oral tradition and cultural complexity, the information is gone forever. It's heartbreaking how often I have come across such careless collection of traditional mnemonic objects, prized for their exotic appeal with no appreciation of their true purpose. There is so much that we could learn from those who understood how to use such devices properly.

As I read *The Roll Call of the Iroquois Chiefs*, and unfolded the diagram that shows all sides of the cane, I was struck by how closely it resembled the genealogy staves of the Māori in design and purpose. I used ideas from both cultures to design my own genealogy staves to help me recall history.

My genealogy staves

I have carved memory staves inspired by those of the Iroquois and Māori. Unfortunately, there is very little information about my family as individuals further back than three or four generations. Instead, I have chosen to record the European Royal families. The British royal family is on the leading face because it is already familiar to me. The relationship to other European families is reflected on the other sides, plus references to major global events.

I have another stave devoted to Chinese dynasties. The ruling houses or clans are on the leading face. On the other sides I have the Japanese and Korean dynasties and rulers indicating the relationship between them politically and culturally. Since creating this stave, a great deal of contemporary politics in that part of the world has made a lot more sense to me because I have some understanding of the background and close relationship between the countries.

I created each stave from lists in books and online. Much of the knowledge became familiar just from the process of designing and carving the stave. I then use the device for revision, which keeps the information fresh. I enjoyed carving the European and Asian staves and then felt guilty because I didn't know our own Australian prime ministers. So they now have a stave as well.

Like the other memory devices, I do not need to have the stave with me to be able to feel its shape in my imagination or see the symbols, although it is obviously much easier when I have it in my hand. My recitations of dynasties and rulers become more song-like every time I perform them. The songs have been added to the ceremonial cycle on the carved wooden balls mentioned above. The more you play with memory devices, the more you will find that integrating them, as indigenous cultures tend to do, can be extremely effective.

Objects acting on a tiny stage

The African Yoruba people memorise vast amounts of information using a method that seemed, when I first heard about it, far too complicated to be realistic. In the simplest version, the Yoruba 'diviner' tosses sixteen cowrie shells on a tray and counts the number of shells that landed mouth up. There are seventeen possible outcomes, from zero to sixteen mouths facing up. Each outcome is associated with a mythological ancestor. These ancestors are ranked from the most senior to the most junior. At each level of this hierarchy, there is associated information.

The song-poetry for this system has been recorded and documented from the memory of one diviner. It's a 305-page work.[1] Having been astounded by this feat, I discovered that there is an even more complex Yoruba system, known as *Ifá* divination, where the oral specialist, the *babaláwo*, tosses sixteen palm nuts and recites verses according to the pattern in which they fall. He then tosses them a second time, adding further information to the reading. Each combination acts as a mnemonic for stories recalling knowledge covering a huge range of scientific, cultural, historic, spiritual, legal and ethical topics. Over 1200 verses have been recorded from the memory of one *babaláwo*.[2]

Of course, I had to test the system. I armed myself with sixteen cowrie shells and a tray on which to toss them.

One genre of information retained by all indigenous cultures is plant classification. My seventeen mythological beings represent each of the plant groupings in our garden, both domestic and wild species. A second toss enables each of those 'beings' to be matched with seventeen further options, giving 289 combinations. That will allow further complexity in the groups that I need to know more about—in some cases down to species level.

I have found that the system works, but it is very difficult and not one I would recommend. The Yoruba must be brilliant.

During my research, I constantly found references to indigenous cultures all over the world, not just the Yoruba, using sets of objects as memory aids when telling stories. Maybe there was a system that suited me better.

The highland Mayan timekeepers arranged seeds on the ground as a mnemonic technique. Australian Aboriginal storytellers, still active, draw in the soil and use leaves, sticks and stones to tell stories on a stage set in the cleared ground. The human brain is a pattern-seeking device. It naturally responds well to the layouts and movement. This seemed like a worthwhile method to try.

Western culture makes so many references to Greek and Roman mythology that I had previously tried a number of methods to memorise the stories. But not only were the gods hugely promiscuous—breeding with immortals, humans and various permutations of living creatures—lots of related mythology was not about the gods themselves. I had failed to come up with any nice neat linear memory technique. So I decided to copy the indigenous practice of storytelling through objects to see how well it worked.

I love the grain of natural wood and the patterns on stones and shells, so telling stories with these objects—with my tabletop as my private stage—worked superbly for me. This was one of the easiest memory methods to implement.

I often start from the beginning of time, but it is possible to jump to any act of the epic performance. To begin, I place my Chaos stone at the top of the stage. Each layout, each stage setting, is fixed in my memory, which enables me to recall relationships at will. I place children below their parents, husbands and wives next to each other. Well, as much as possible given the mayhem of matings.

I move my stones, shells and wooden hearts around the table to act out scenes. Husbands have a home base next to their wives, but they jump across the table when they play away. Zeus does an awful lot of jumping. When Zeus dies, I take his wooden heart (I do like that phrase) from the stage.

I can use the same wooden heart for another character later in the action by turning it over. The changing wood grain makes it look so different that it has no further connection to Zeus. If I spin it so that the point now faces away from me when Zeus had it facing towards me, then it looks quite different again. That way, I don't need as many objects as I have characters in the stories. Those who are not in play rest on the side of my stage, waiting in the wings until they are needed. By the end of the first story, I have a setting like that shown in Plate 14.

Bit by bit, the acts of the play were added and the movements settled into their permanent choreography. I found that the rhythm of the story evolved alongside the rhythm of the movement of my hands as I rearranged the props. I have developed some rather melodramatic hand gestures to aid my narration. Any time I want, my memory will replay the stylised movements. I perform a dance with my hands almost without thinking about it. This reflects what musicians and actors have told me about muscle memory. (More on that in Chapter 8.)

My purpose for this specific memory task was not to be able to reel off all the stories in order and give hours of performances for anyone else. The purpose was for me to be able to think about any character and recall his or her story in their mythological context.

A memory method employing performances on a miniature stage with objects is a particularly strong method when your information will not fit into a neat linear format. Obviously, this is great for historical narratives, but I wanted to see if it would work for more abstract concepts.

I chose to act out the radioactive decay chains for uranium, thorium, neptunium and actinium with the same objects. This just required one more step—giving the elements a personality to link to the objects. Plutonium was easy; there was a dog-like shape in the grain in one of the wooden hearts, so I thought of the canine cartoon character Pluto. Some of the elements are named for Greek gods. For example, uranium was named for the planet that was named for the god Uranus. My Uranus stone became my uranium element. Neptune was the Roman god of the sea. His equivalent in Greek mythology was Poseidon, so my Poseidon wooden heart served for the element neptunium. Others, like radon, bismuth and thallium, took a bit more work. I soon had them all represented and danced my hands around as the elements decayed into each other. I turned them around as they changed isotopes, so U–238 is the same stone but at a different angle to U–235. Those that became stable, such as lead when it reached Pb–208, I placed my hand firmly on so they could dance no more.

I had previously included the decay series in the memory palace for the periodic table but found this performance method worked much better. I didn't need to remember the atomic number of any element; the memory palace did that. The two memory methods meshed together seamlessly.

The more you play with a range of memory methods, the more you will be able to detect which works best for the data you are trying to encode. Then they will just serve to reinforce each other.

The memory device that never leaves: your body

There is one memory device that you have with you the whole time, apart from your brain, of course: your body.

Aboriginal Elders are referred to as Uncle or Auntie as a term of respect. Auntie Julie McHale described to me the way the Dja Dja Wurrung, on whose country I live, use body parts as a counting method. Body tally systems are known around the world. Although they vary, they usually start at the little finger of one hand and progress right around the body, ending up at the little finger of the other hand. It then gets more complex, using repeat cycles or other methods for higher numbers. There is also evidence of much more complex mathematics.

I have often heard it said that indigenous cultures don't count. That misconception could have arisen because in some cultures there weren't separate words for numbers. Imagine counting a flock of birds or a herd of animals. As you note each one, you touch body parts in order. Let's say that using this process 26 animals lands you on your left shin.

So the word for 26 and the word for left shin would be the same, the context telling you whether you're talking about a tally or a body part. Along comes an ethnographer and asks you the word for the number 26. You give the word for left shin. In the vastly superior way that typified European interactions with indigenous cultures for most of history, they note down that you have no mathematics because you don't even have words for numbers. They investigate no further.

Are you more likely to remember a body part you pinched hard or an abstract number? I'm all for body tallying to count low numbers.

You can also use your body as a memory palace. All you need to do is take a trip around your body and consider each part as a memory location. It is a great memory journey for things you want to contemplate often. This is the order of body parts I use:

1. Right hand
2. Right wrist
3. Right forearm
4. Right elbow
5. Right biceps
6. Right shoulder
7. Right ear
8. Top of my head
9. Eyes
10. Nose
11. Mouth
12. Chin
13. Left ear
14. Left shoulder
15. Left biceps
16. Left elbow
17. Left forearm
18. Left wrist
19. Left hand
20. Neck
21. Chest
22. Stomach
23. Backside
24. Left thigh
25. Left knee
26. Left shin

27. Left ankle
28. Left foot
29. Left toes, *which kick my* . . .
30. Right thigh
31. Right knee
32. Right shin
33. Right ankle
34. Right foot
35. Right toes

There are freckles and spots and wrinkles and lines all over the place, should you want to get more complex.

Fundamental to my current research is the way different writers, from classical times until the present, wrote about memory methods. I want to ponder the way the implementation and purpose changed with time by recalling their writings. So I mentally move around my body, contemplating each of the memory techniques and the people who promoted them. Unlike my 'ancestors' described in the previous chapter, these are people of consequence to my research, but not ones I intend to study in depth as individuals. It is what they *talked* about that matters.

I review my body palace (isn't that a lovely combination of words?) when I am bored—when I'm listening to dull speeches or in meetings or during any of those annoying times when I have to sit still and wait for something more interesting to occur. I discreetly wiggle and squeeze, and tense and relax various parts. I am often reminded of an experiment or person I have not contemplated for a while.

Astronomy in the palm of my hands

I have no doubt at all that the palms of hands would have been used as a memory device throughout time. They are too perfect for the task for it to be otherwise. But my search for evidence of this had just led me to endless articles about palmistry, so I was unduly excited when I saw a huge hand on the wall of the State Library of Victoria's exhibition, *World of the Book*. It was labelled as an illustration from a devotional text called *Schatzbehalter* written by Stephan Fridolin (c. 1430–1498), an 'Observant Franciscan friar'. The little explanatory sign, which pleased me so greatly, described the hand as a mnemonic device for giving speeches. I knew it!

Further research revealed that the *Schatzbehalter* hands are marked with numbers that refer to individual meditations given in the book, enabling readers to memorise 100 themes (see Plate 15).

I then discovered the Guidonian Hand method, which was used in medieval times to assist singers learning to sight-sing medieval music. The eleventh-century monk Guido of Arezzo created a hexachord system spanning almost three octaves. The diagram of the hand acted as a memory aid for practising his system, with each joint representing a note.

I decided to test my hands as a mnemonic device for something more permanent than a speech, given I could use my visual alphabet for that purpose. So I encoded astronomy topics to the fronts and backs of my hands. Can I resist the obvious comment on the saying 'I know it like the back of my hand'? No, I can't. I didn't know the back of my hands at all. And my palms are even more lively.

I used the palm and back of my left hand for the science and my right hand for the history of astronomy, as shown in Figure 5.4.

It works the way any memory palace works: I make some kind of imaginative connection between the physical location

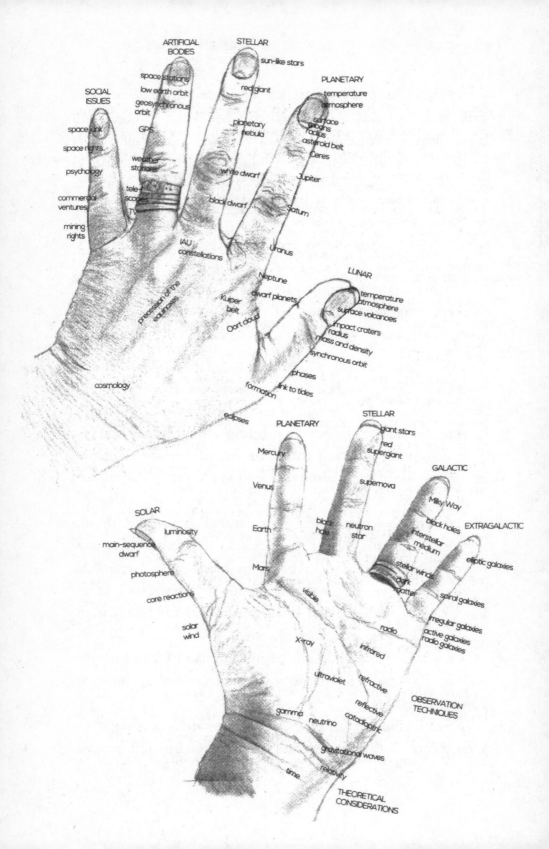

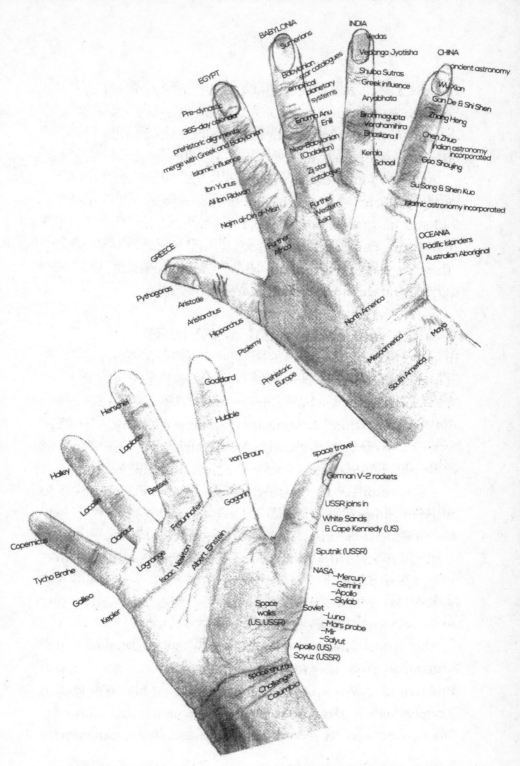

FIGURE 5.4 Using my hands as a mnemonic for astronomy—every feature on both left and right, back and front, can be utilised. (LYNNE KELLY)

and the knowledge to be remembered, and then encode the knowledge in an image or story. It is not going to work for extremely complex information. Like all memory palaces, it will provide a grounding and structure for higher levels of learning.

At first I found the abstract nature of my hand a bit challenging. Then I realised that my fingers jiggle and point and fold and join forces, and my palms clap and wave and move at all sorts of angles. I can pinch bits and scratch them, while others are out of reach of my nails. Adding in movement and touch made the stories much more memorable.

Wearing your memory aids as jewellery

Belts, beads and knotted cords have been used as memory devices by indigenous people around the world, and throughout time. Why not accessorise your memory aids? You can be practising and recalling while standing in a queue, or waiting for your coffee to be made or the kettle to boil, during the ads on TV or while pretending to be interested in the office meeting.

I was inspired to wear my memory aids by a number of different indigenous traditions. I saw examples of the beaded *wampum* in museums. The skill to make them and their aesthetic appeal left a profound impression. I cannot imagine how much more they must mean to the Native American Eastern Woodlands tribes who use beaded belts and other items as sacred and pragmatic memory devices.

The word *'wampum'* refers to small purple or white shell beads made from the quahog shell. The beads were threaded into strings or woven into belts. Leading individuals would own *wampum*, which demonstrated their credentials and authority. They were used to record oral tradition, to commemorate treaties between Nations, to list laws, for messaging and as prompts when narrating teachings. Both the design and the number of beads were significant in recording knowledge.

Why not make your own belts, necklaces and bracelets by weaving the objects in a pattern to match the knowledge you wish to encode? I am encoding a study of the plays of Shakespeare. I have one bead per play in chronological order threaded onto a necklace. The plays are usually divided into comedies, tragedies and histories, so I have different colours and styles of bead for each of these categories.

The order in which Shakespeare is thought to have written the plays forms the order of the beads, so the pattern on my necklace was dictated by Shakespeare himself. Probably. That does assume that the historians have the chronology right, which is by no means certain. But I like to think that Shakespeare designed my necklace and I am unlikely to be convinced otherwise. My string of beads for Shakespeare's plays, and for *A Midsummer Night's Dream*, can be seen in Plate 16.

Mostly, I chose beads with patterns on them. If I want to add details, I use permanent markers to add dots and lines and splotches on the beads in black and gold and silver. I separate the beads with a small bead just to make it easier to distinguish them.

I have strings of beads for the story of each play, which I am now stitching to a soft cloth handbag. I could wear them as necklaces, bracelets, decorations on clothing or combined as a belt, but I have chosen to decorate a handbag. In this wonderfully enlightened time, men can wear jewellery just as women do, and either gender can keep their beaded delights in pockets and handbags.

The string for *A Midsummer Night's Dream* features nineteen distinctive beads for the main characters. The little brown beads are just separators. Starting from the left are the humans, represented as a group by the cowrie shell. The individual characters each have a fancy bead. Following are the workmen, represented as a group by the larger buff-coloured wooden wheel. The individual workmen each have a different-coloured wooden wheel–shaped button. The fairies are represented by strangely shaped beads made of pinkish or brownish shell, each different enough to represent the fairy characters individually. The time I took to choose these beads embedded each character in my mind.

There are five acts, the first four having two scenes and the final act having only one. I represented the acts with long thin brown beads. The scenes are thin white beads. So once the characters have been introduced from the left of the necklace, I started encoding the acts and scenes. Act 1, Scene 1, features the humans, so following the single long thin brown bead and the single thin white bead is a cowrie shell, telling me that the scene involves only the humans. Then come two white scene beads and a wooden wheel, telling me that Scene 2 is all about the workmen. Act 2, Scene 2, has fairies and humans, so following the two scene shells is both a fairy and a cowrie shell.

As I recall the play from the string, I know if I have messed up the sequence or action if I have workmen prancing around in a scene that contains only fairies and humans.

I have discovered so many gorgeous beads since I started this project. And I am getting much more ambitious in my designs. I really look forward to beading something akin to the *wampum* belts rather than just stringing beads on a single thread. Sitting on my desk are beading books with the most seductive selection of designs, all of which could be manipulated to suit information encoding. I long for the time to bead to my 'heart's content', as Shakespeare said in *Henry VI, Part II* and *The Merchant of Venice*. I wonder if I could encode every phrase coined by Shakespeare in beading?

Students are allowed to wear jewellery into exams, aren't they? Would that be cheating? Of course, like most of the memory devices, repeated use will commit them to memory along with their encoded record, so you wouldn't need to wear the jewellery. As you had worked to encode and memorise it, why would it be banned?

You will find that, while you could, you don't *need* to carry around the devices talked about in this chapter so far. You will soon memorise them and use them entirely through your imagination. If you sketch them a few times, that process will be even faster. But although the object may not be needed anymore, your affection for it will not diminish. While worthless to anyone else, my memory devices are incredibly precious to me.

Knot your strings into a personal *khipu*

I haven't found my knotted cord devices quite as memorable. Tying knots on a piece of string is a technique used by multiple cultures, literate and non-literate, around the world. Many Native American tribes, for example, used knotted strings to note the number of days until a meeting, discarding a knot each day until the time of the meeting.

By far the most famous is the intricately knotted *khipu* of the Inca, as shown in Plate 17. The *khipu* (also spelt *quipu*) was used by Andean knowledge specialists, the *Quipucamayos*. They held the main cord horizontally and let the attached cords hang vertically. These secondary cords could number anywhere from dozens to hundreds.

Researchers initially thought that *khipu* were used only for accounting records. They now realise that they were also used as a memory aid for narratives, laws, rituals and histories, demographic data, tributes, a form of calendar and probably a lot more. The *khipu*—along with memory paths, the *ceques*— enabled the Inca to create the most extensive non-literate culture in the Americas, if not the world. But is 'non-literate' the right term for the Inca?

Between the establishment of the South American empire by Pachacuti Inca Yupanqui in 1438, to the execution of the last Inca, Atahualpa, by the Spanish in 1533, the Incan empire flourished. The Mesoamerican Aztecs and Maya, just to their north, had written scripts. The Inca did not, but there's a great deal of debate as to whether the *khipu* is purely a mnemonic device or if it is in fact much closer to writing. The answer, if there is to be academic agreement, is probably a few decades away.

I simply adore these extraordinary objects, which can be seen in many museums. They are often laid out to look like necklaces, although I have read nothing that suggests that they were worn that way. I gather they may have been slung across the body. I wear mine as a belt, hanging shawl-like over a long skirt.

I am using my version of a *khipu* to record the history of Western art. Each string represents an artist. They hang from the primary cord in chronological order. Harvard professor

Gary Urton describes all the variations you can use for the knots: the number, spacing, direction of the twists, colours, feel and ply of the cords.[3] All these traits can act to encode the information. I am using every variation he suggests to encode the initials of the artist, their dates and a code for the type of art: landscape, portrait, abstract, sculpture and so on.

The beauty of using knotted cord is that if you need a new category, then you can just come up with a new form of knot, or knots, to indicate it. If you want to add a new cord, you can hang extras from the secondary cords, as the Inca did. And, I am delighted to report from extensive experience, you can re-knot them when you've messed up.

Making my *khipu* was wonderful. I loved buying a range of cords and choosing the colours and carefully threading and knotting it. I love that it is such an amazingly adaptable device that I can keep changing it and updating it and adding new information. That meant that I could start straightaway and then change my mind about the coding as I grew to understand what would work best. I had quite a few false starts, but all that did was force me to engage with the information more deeply. I was learning about art at an astonishing rate, just by making the *khipu*. I found I wanted a sub-*khipu* just for Australian art, as that is what I see most in the galleries here. So I made one. I wonder if the Inca had sub-*khipus*.

Although my *khipu* is by far the most adaptable of all the portable memory devices, I have also found it less memorable than any of the other devices described above. I wonder if they were ever intended to be fully memorised, which is fine because *khipus* are very light and take up little space so why *not* just carry it with you? From experience, I do recommend binding them neatly or you will get to spend many hours untangling them.

So what is the difference between memory devices and writing? It wasn't a simple matter of the sudden invention of writing and the dumping of all memory aids. Life is never that simple.

CHAPTER 6

When art becomes writing

I adore my second *lukasa*. My *lukasa* for birds worked wonderfully well but I knew that I was not taking full advantage of the lesson from the Luba experts in terms of the shape and design. I had held the genuine article at the Brooklyn Museum in New York—so Tom Chippindall replicated the style for me as he had done with the carved wooden balls. My second *lukasa* had the same weight and shape as the real thing, but the beads and shells were arranged to suit my needs. As Tom drank his tea and ate his chocolate cake reward, I fell in love with the way this *lukasa* felt. It was so light, and curved to fit my hand. Tom had even added a tiny bead representation of a *khipu* in exactly the right place. I was delighted.

Ironic, I know, but I have encoded the story of writing to my second *lukasa*. Tom had placed the beads in a general shape representing the knowledge I wanted to encode. He placed enough beads and shells to enable me to expand my associations as I learned the details. The Luba would have already known their knowledge when they made their *lukasas*, and their design would have evolved over a very long time alongside their performances. I didn't have that advantage. Two real *lukasas* and my two *lukasas* are shown in Plate 13.

This chapter is not a complete history of the evolution of writing—that would take many chapters, many books—but it gives a foundation on which to build. As you read on, you'll be able to see the way my second *lukasa* was designed to tell this story. Figure 6.1 is an annotated image of it.

When and what was the first writing?

I could pontificate upon the many debates around the definition of writing, but I won't. I would soon get bored and I suspect you would be even more so. The generally accepted definition is that writing is speech converted to some visible form. There is at least some representation of sounds. Writing means that any reader who knows the language can reconstruct exactly what is to be spoken, word for word. Nothing needs to be memorised except how to read. Writing is a mnemonic device.

Using that definition, I doubt that the *khipu*, as described in the previous chapter, is true writing. I suspect that it is a really advanced memory aid that has developed to be as close to writing as a non-literate device can be.

On my *lukasa*, theoretical concepts are represented by the beads across the top. Non-literate memory systems from the previous chapters take up most of the top half of the *lukasa*. The middle carvings represent undeciphered sets of symbols,

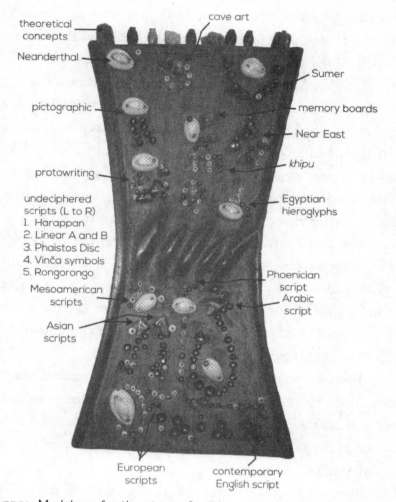

FIGURE 6.1 My *lukasa* for the story of writing, made by Tom Chippindall. (LYNNE KELLY)

such as the Rongorongo glyphs from Easter Island, which may or may not be writing. The lower half is all about writing.

The start of the art-to-writing story

The first shell on the very top left-hand side of the *lukasa* represents formations and markings made by Neanderthals.

Did they practise memory methods? I'd love to claim they did, but it would be pretty speculative based on the evidence to date. Along the top edge, the Neanderthals are followed by the most famous cave art sites, in chronological order.

The earliest cave art has been dated to over 40,000 years ago, although I predict that as dating methods improve we will find deliberate imagery that is much older. In parts of Australia, we have a continuous art record dating back tens of thousands of years. Although the art, along with the society, was gradually changing, there was no break in the art style. We know that Australian Aboriginal cultures use art as an aid to memory because they have told us so. It's a combination of representational and abstract imagery. A great deal is abstract because abstract symbols allow much more complexity to be encoded—a given symbol can represent a multitude of different meanings depending on the context.

Indigenous cultures throughout time have created permanent art on trees and rocks, posts and poles. Decorated posts are particularly effective memory aids, the best known being the totem poles of the north-west coast Native American and Canadian First Nations people.

I have decorated a post on the corner of my studio verandah to encode the 139 species of mammals in my state. They are sequenced in taxonomic order down the pole. I also encoded a classification of the kingdom Animalia to natural stones in our front wall. Stones have so many shapes in their texture that if you take the time to stare at them you will see any animal you want. I'm not going to write in detail about the process because not everyone will have a paintable pole, circle of standing stones or rock wall readily available, and I think the primary value of such large mnemonic art is when it serves a community. Nevertheless, I wanted to test the effectiveness of decorated posts and standing stones because in previous books I have talked about their role in memory systems in the archaeological context. So many ancient monuments from non-literate cultures are constructed ceremonial sites, with standing stones and posts in clearly defined sequences, perfect for large-scale memory palaces.

But these days we're a little tighter on space, so I'd tend to use a standard memory palace or memory board to study mammals and animal classification rather than construct a personal Stonehenge in my garden.

Tibetan mandalas as a memory palace

Memory boards are just one form of small-scale memory palace. Tibetan mandalas are stunning objects that also function as memory devices. You can see just one example, the mandala of Vishnu, in Plate 18. The word 'mandala' usually refers to a circular representation of the cosmos for Hindus and Buddhists, but contemporary artists use it to mean almost any circular geometric pattern that in some way represents a spiritual or metaphysical concept of the universe.

In 2009, I was in New York at the Rubin Museum of Art. It houses one of the largest collections of Himalayan and Tibetan art in North America. The hours seemed so short as I indulged in the stunning exhibitions across the quiet and spacious six floors of galleries.

Jeff Watt, the founding curator and the leading scholar at the Rubin Museum for ten years, is an expert on Himalayan tantric art. This style of art, which is over a thousand years old, continued to develop even after the texts were written down between the fourth and tenth centuries; it enhances literacy, rather than being replaced by it. The majority of works represent Buddha, his followers and their teachings, both spiritual and mundane. As Watt explains: 'Tantric art actually compresses all of the important points of the Buddhist teachings into a very tight visual package of symbols.'[1]

Watt was asked, 'Are the mandala paintings essentially mnemonic devices?'

Yes. The mandala represents the totality of the universe, and each deity resides within one. The relationship between the deity and the mandala is the relationship between the animate and the inanimate. The deity represents oneself and the surrounding mandala represents the world one lives in. Typically you have a figure in a square, representing a deity in his or her palace, with four doors. Tantra's foundation comprises the fundamental Buddhist teachings, so different features of the palace represent what in Buddhism are known as the 'thirty-seven factors of enlightenment'—the steps leading to enlightenment, including and beginning with the Four Noble Truths, which the four doors represent. It's a little bit like Shakespeare's Globe Theatre, in which actors used architectural features to remember their lines. Likewise, the tantric practitioner uses the different features of a mandala— a metaphorical palace—to remember the thirty-seven factors of enlightenment.

From Shakespeare to Tibetan deities—how could I resist trying it?

My mandalas for science and law

The mandala format really suits a topic where a character is central to the teaching, but it also works if there is a central *theme*. I chose two examples from very different fields, neither as ambitious as the mandala of Vishnu. I adjusted the number of little spaces in my rough plans to suit the number of points in my notes.

There are many images of mandalas online to admire and use as an inspiration for your own design. And all the while you are thinking about your plan, you are becoming more engaged with the specific knowledge. The process is as important as the final product, if not more so.

I chose two diverse topics to test the effectiveness of mandalas as memory aids: Young's double-slit experiment for physics and the Eddie Mabo case for law.

I chose Young's experiment because it had a cosmological, almost spiritual impact on me many years ago as a physics student. The double-slit experiment was first performed in 1801, and the ramifications and implications of this experiment still affect major questions in physics today. Originally, the question being asked was whether light is made of small particles or whether it is a wave.

This debate pitted Thomas Young (a believer in the wave theory) against the dominant figure of Isaac Newton (a believer in light particles). Subsequent double-slit experiments have demonstrated that light is simply a confusing entity and headed us into quantum mechanics. Consequently, my mandala, starting with Thomas Young and Isaac Newton, has a degree of complexity that can only be fully explored with layer upon layer of study.

The mandala worked better for this information than a linear design, such as a memory palace, because the ideas that originate from Young's experiment lead in many different directions. I represented these around the circle with the relevant characters. I also included empty panels for ideas I didn't yet understand (a lot of quantum mechanics), ready for expansion. So much has happened since Thomas Young's time with double-slit experiments; my mandala inspires me to find out more.

I wanted to try another context for a mandala; studying law requires a great deal of memory work, so I chose a historically significant legal case.

Before the Eddie Mabo case, Australian Indigenous people didn't have rights to the land they had occupied for tens of thousands of years. The British arrived a mere 200 years ago and claimed it for England on the grounds that it was unoccupied land, *terra nullius*. The Queensland Government opposed giving land rights to Aboriginal people. The High Court eventually ruled in favour of Eddie Mabo and his Meriam People of the Torres Strait, a decision that led to land rights for the hundreds of Indigenous language groups across the continent and associated islands. Sadly, the case dragged on for so long that Eddie Mabo did not live to hear the decision.

In my mandala, I have the major players battling and supporting Mabo, left against right. At the top is the history of the law as it stood before Mabo. At the bottom are some panels about the subsequent *Native Title Act* and land rights rulings.

As I created each tiny drawing, I concentrated on the facts I was representing. Transforming each part of the story into a visual representation forced me to ensure that I understood.

Without that understanding it would have been impossible to know exactly where to place each image in my mandala.

That's why diagrams that organise information and adorn it with lots of shapes and colours are so effective. As soon as I started designing my first mandala, I was reminded of the enthusiasm in educational circles for mind maps, as they are widely known. They are another way of creating a memory palace on paper. Like the mandalas, they use a central image for the main topic, branching out with colourful subheadings in all directions.

I found the most powerful understanding came while roughly sketching the mandalas, as I pondered the stories of Young and Mabo, of the people and their impact. My finished product was just an added bonus.

This is no radical discovery of mine. The intricate and beautiful Tibetan Buddhist sand mandalas are created from coloured sand and then ritually dismantled in ceremony. Temporary art is a feature of indigenous cultures the world over. I have admired many Australian Aboriginal and Native American drawings in soil and sand, only to see them wiped away. They were never intended to be permanent. It is the process of their construction that serves to make them memorable and conveys the teachings of the stories.

Are they mnemonic symbols or are they writing?

We're now down to the middle of my *lukasa*. The carved ridges are there to remind me of ancient objects adorned with symbols and signs that look as though they were structured to be read. But were they actually writing? I can add more examples into the dips and grain around the ridges as I come across them, but at this stage my *lukasa* reminds me of the five most famous scripts that have to date defied deciphering.

The first dates from between 3500 and 1900 BCE, a set of symbols known as the Indus or Harappan script used in what is now India. It was developed by the Indus Valley civilisation. Although there are over 4000 samples, they are all short segments.

The second is two sets of hieroglyphs found on ancient Greek artefacts. Linear A, from around 1700 BCE, has yet to be deciphered while Linear B, from about 1375 BCE, has been deciphered.

The third is the Phaistos Disc—shown in Figure 6.2, should you be inspired to solve one of the world's great mysteries—which may be a hoax, or may be a fired-clay artefact from the

FIGURE 6.2 The Phaistos Disc, which may, or may not, be a script waiting for someone to decipher, date and verify. (C. MESSIER)

Minoan palace of Phaistos on Crete, dating to somewhere in the few centuries before 1500 BCE. The disc measures sixteen centimetres in diameter and is stamped on both sides with mysterious symbols. No other examples like this have ever been found.

Also intriguing are the relics from the Neolithic Vinča culture of central and south-eastern Europe, the fourth example. These symbols, created between the sixth and fifth millennia BCE, became increasingly complex and look like a proper written script, but there is much debate among archae-ologists if they are really writing or some form of mnemonics.

The fifth example is from Rapa Nui, also known as Easter Island, where over twenty wooden objects were found covered in symbols that certainly look like writing and is known as Rongorongo. The oral tradition says that only a small elite could ever read these sacred objects. Is this a written script that originated independently on this remote island? Unpreserved wood doesn't survive well, so very old samples will have been lost. The oldest known sample is dated to 1851, well over a hundred years after the first European contact. So is this a case of the idea of writing being adopted by the indigenous islanders observing the early European visitors? It wouldn't be unique if it were.

Shong Lue Yang, a twentieth-century Hmong spiritual leader living in Laos, created a semi-syllabic Pahawh script for dialects of the Hmong and Khmu languages. He was an illiterate subsistence farmer and basket maker who would have only observed writing but, beginning in 1959, he created his own script. Similarly, Sequoyah, a Cherokee silversmith born around 1770, saw white settlers using writing and created his own script for the Cherokee language. It took him over a decade but eventually he completed a Cherokee syllabary, a set of symbols representing every syllable. He then managed to convince the Cherokee leaders to use the script. Soon the Cherokee Nation literacy rate surpassed that of the non-indigenous settlers.

The Cherokee script was not the first developed in the Americas. Unfortunately, the aesthetically gorgeous Meso-american scripts could not survive the onslaught of the Spanish invasion.

In the pre-Columbian New World, many cultures had scripts, some of which had phonetic elements, possibly dating back over 3000 years. Zapotec and Olmec writing are considered the earliest but are not fully understood. The earliest deciphered script is that of the Maya, a culture dating back over 2000 years. Around 300 BCE, they developed their own written script, including symbols for syllables. They adorned stone pillars and wrote in codices for 1200 years. The tragedy is that they could not maintain their script right through their long history. Starting in 1521, the Spanish came and destroyed it all. A century later, no one who could read the beautiful symbols remained, even though there are still active Mayan communities today. It is only in the last 50 years that academics have managed to decipher the symbols so that most ancient Mayan writing can now be read.

Humanity's first symbols were pictographs representing objects or ideas—it is a massive leap in human creativity to jump to a system with symbols that represent sounds. Many of the early writing systems included symbols for syllables and some created signs for every elementary sound. Some writing systems, such as Chinese, created complex graphics, which included symbols for meaning and sound within a single character.

From art to writing in China

One of the reasons I love learning Chinese is the sense of history: it is the only language spoken today that can be traced back to its preliterate pictographic past. It is a totally different way of writing a language from English, let alone speaking it. China boasts a calligraphic art tradition that has me mesmerised. I encode all the information about the pre-literate symbols found in China to a sequence of little brown beads topped with a shell on the left-hand side of the bottom half of my *lukasa*.

Around 6600 BCE, symbols were carved on tortoise shells found at Henan in central China. Known as the Jiahu symbols, they are not currently thought to be systematic writing so much as mnemonic markings; however, these symbols demonstrate that Chinese signs for language, in some form, date back to the Neolithic Peiligang culture at least.

Cliff carvings at Damaidi in north-central China date back to about 6000 BCE. There are 8453 different characters. These have been interpreted as depicting social, hunting and agricultural scenes, as well as the sun, moon, stars and gods. If either the Jiahu symbols or Damaidi characters represent genuine writing, then the Chinese beat any other culture to a script by thousands of years.

However, the general agreement is that the first script is that of the Sumerians, dating from around 3000 BCE. The purveyors of theories that are accepted as 'general agreement' don't like giving up the pedestal without a massive fight. I'll leave all that to the academics.

The Asian scripts are represented by stones and beads flowing down the left-hand side. The tiny polished stones represent mandalas, and handscrolls, which I will describe below. The descending beads enable a chronology, from ancient times to the contemporary Chinese script, and those which are related to it, such as Korean and Japanese characters.

In northern China, oracle bones were found adorned with what is now considered to be the precursors of the contemporary Chinese script. These date from around 1200 BCE, the Shang Dynasty. They still looked like pictures but are considered the first example of what is definitely Chinese writing.

I'm not greatly concerned about who won the race to a script, or whether one got the idea from the other while bartering over goods. What I find so beautiful is that there is a continuous pattern from the 8000-year-old tortoise shells to the cliffs to the bones all the way to the symbols my Chinese friends are teaching me today. How amazing is that?

Sometime between the tortoise shells and the bones, the concept of symbols representing the sounds of syllables was introduced. Around 200 BCE, 'the first emperor' Qin Shi Huang united the massive country and all its local cultures. He made rules about how to write each character. Over thousands of years the writing became more stylised, incorporating signs for both sound and meaning in the same character. In 1956, many of the characters were simplified by government rules, reducing the number of strokes and creating a new standardised character set.

Alongside the Chinese characters are a multitude of art forms that serve to prompt memory and narrative. On two visits to the United States, I spent hours with the Chinese narrative scrolls at the Metropolitan Museum of Art in New York. I was totally besotted with their beauty and am convinced this form of art can be used as a contemporary memory aid, especially for topics that have a linear progression, such as history.

Zhang Hongxing writes about handscroll paintings in his introduction to the stunningly beautiful book *Masterpieces of Chinese Painting, 700–1900*. He describes the way hand-scrolls can display a whole narrative in a single long painting. An actor can reappear in scene after scene, as if on a painted stage, moving the play through time or across the landscape, or both. It is this sequential action that makes the format such a fantastic mnemonic technology.

Zhang talks about a scroll dated to around 344–406 CE, *The Goddess of the Luo River Ode* attributed to Gu Kaizhi:

In a late Northern Song copy of this masterly scroll, the episodes of the story of the poet's ill-fated romance with the river goddess unfold through a continuous landscape, and the images of the two lovers and their attendants appear again and again. It is unclear whether Chinese painters developed this technique of continuous and linear pictorial narration of classical subjects over the first, fifth and sixth centuries independently of the narrative tradition in Indian Buddhist art, which was being introduced into China during this period, or whether they adapted important elements from the illustrated Buddhist texts originating in India to develop techniques for illustrating classical Chinese subjects . . . In any case, from the 11th century on, painters worked to explore temporal and spatial progression in the horizontal format not just with classical poetry and Buddhist or Confucian tales, but also with landscape and genre scenes. For many artists, a continuous pictorial progression was important because they felt that the underlying purpose in painting such subjects was to provide the viewer with an imaginary travelling experience.[2]

I was convinced that I needed to trot off to New York again if I hoped to see such beautiful objects again. I was not yet aware they were not a uniquely Chinese tradition, so was not searching among the exhibits of other cultures. In the National Gallery of Victoria, only a few hours from home, there is a Japanese narrative scroll filling the length of a twelve-metre case, as shown in Plate 19. Narrative scrolls were produced in Korea and India as well. The *patachitra* scroll artists of West Bengal in East India were traditionally wandering minstrels singing the epics and revealing the scrolls as they performed. This is still a practised art form, with *patachitra* scrolls representing contemporary events from the French Revolution to the 9/11 attacks. More often, they portray stories from the great Hindu epics, including the Ramayana.

Narrative scrolls are not just an Asian tradition: the Bayeux Tapestry is a 70-metre long embroidered cloth. The narrative covers the events leading up to the Norman conquest of England and culminating in the Battle of Hastings in 1066. Despite the name, it was made in England and probably embroidered within a few years of the conclusion of the story it tells.

Although the styles varied over time and between cultures and artists, the fundamental format remained the same:

education taught through a stunning beauty. I was both daunted and excited to attempt my own version of a narrative scroll.

My narrative scroll: the story of timekeeping

I am convinced that a contemporary form of narrative scroll could be a wonderful memory device. I know from experience in the classroom that changing the way you represent the knowledge—known as a modality shift—makes you remember information far better than just reading it over and over. You can read without really concentrating, but if you have to change written information into images, or songs to mind maps, you will remember it. To change the modality, you are forced to focus on the information.

Your narrative scrolls need not be time-consuming artistic endeavours. They could be quick sketches. It is the *process* that will commit the knowledge to memory.

I chose to tell the story of timekeeping through the ages in my handscroll. My little rapscallion watches the changes of the seasons, reckons by the stars, puts sticks in the ground as a sundial, swings from the pendulum of the grandfather clock and jumps in fear when tiny doors spring open to issue forth a thunderous cuckoo. I added background images to indicate the location when appropriate. Adding characters to create a mythology made my chronology of timekeeping so much more memorable—and so much prettier!

From Sumer to the world

In the top right of my writing *lukasa* is a shell that represents the Sumerians. Beads descend and eventually split into the Western languages, on the bottom left, or western, side of

159

the board, and the Near Eastern languages (as long as you use a broad enough definition) on the right, or eastern, side. Along the bottom are the scripts that led to our contemporary English writing. They are given more prominence that would be justifiable usually, but I love calligraphy so I wanted them there—this is my own *lukasa* and so expresses my personal biases.

The Sumerians built their city-states around the delta of the Tigris and Euphrates rivers in what is now southern Iraq. What we consider writing didn't really start until thousands of years after the first Sumerian symbols appeared. Early farmers started using tokens marked with simple images to label products for trade about 9000 years ago. People settled. Cities grew. Manufacturing led to more specialists and trade, which led to more complex tokens. Later still, a blunted reed was used to make impressions, leading to a wedge-shaped writing known as cuneiform.

Early cuneiform could only record numbers of bags of grain and animal stock but they could be read in any language. Between 3500 and 3000 BCE, the Sumerians started gradually representing sounds. Words that were hard to represent with images were represented by objects that sounded similar. Now only somebody literate in the Sumerian language could read the clay tablets. Sound had entered the symbolic world, but it made the writing only readable to those who sounded the same, to those who spoke the same language.

The trend to use images for sounds as well as objects continued to grow until every utterance that Sumerians made could be represented by symbols. Writing in cuneiform became so popular that over 30,000 cuneiform tablets dating from around 700 BCE were found at just one archaeological site, the Royal Library of Ashurbanipal in what is now northern Iraq.

Neighbours in western Asia recognised the value of writing so soon followed: the Cretans and Elamites each developed their own scripts, as did the people of the Indus Valley, now in Pakistan, as well as the Egyptians.

Writing appeared very suddenly in Egypt, sometime around 3000 BCE. Egyptian hieroglyphs were mostly phonetic, with some pictographs thrown in to help. Because they were phonetic from the start, it seems almost certain that the Egyptians copied the idea from the Sumerians rather than developing it themselves.

Every time I have visited, the biggest crowd in the British Museum is staring at the Rosetta Stone, shown in Plate 20. The discovery of the stone in 1799 allowed the deciphering of Egyptian hieroglyphics: it is a political decree, written in three languages, from 196 BCE. The top is written in the ancient Egyptian hieroglyphic script. The middle is another Ancient Egyptian writing known as Demotic. And, incredibly usefully, the bottom is in Ancient Greek, which gave the key to deciphering the Egyptian hieroglyphs.

The Phoenicians based their alphabet on the Egyptian hieroglyphs. They got rid of the pictographs and went entirely for sound. They created an alphabet consisting of 22 letters, all consonants. Gradually they started using some of the consonants for vowel sounds as well. The alphabet was spread through trade right across the Mediterranean world. Soon there were Aramaic and Greek alphabets and then writing systems from western Asia to Africa and Europe. The Greeks added symbols specifically for vowel sounds.

But just because there was writing doesn't mean that all the old memory methods were lost. They hung around in Europe for thousands of years, through Classical antiquity, the Middle Ages and right into the Renaissance.

Lessons from Greco-Roman times

The most famous of all the pre-literate Greek orators was Homer, whom some consider as performing the role of an encyclopedia for his society, storing and conveying all the knowledge needed by the general population. No one knows for sure, but the *Iliad* and the *Odyssey* are thought to date to around 800 BCE. Some scholars represent Homer as a blind, illiterate genius poet who singlehandedly created the two great epics. Others doubt he existed at all.

The Greek epics, as performed by Homeric orators, were essentially massive repositories of all the cultural information of the society: customs, laws, social proprieties, history, geography, science, technology and knowledge of far-away places. The poems were both educational and entertaining. Without the information, the culture could not survive. Without the entertainment, people would not be willing to listen in order to receive, and remain familiar with, the encoded knowledge. Sound familiar? This blend is exactly what I found in the performances of indigenous cultures from around the world.

Although there was some form of writing during his time, it was recorded that Homeric bards could recite the entire *Iliad* from memory, nearly 16,000 lines. Experts estimate that it would have taken at least four long evening sessions. There are nearly as many lines again in the *Odyssey*. The epics were written down between 750 and 650 BCE.

The *Iliad* and the *Odyssey* were composed to be performed, not read. They have a repeated beat. They have a sequenced story, with the setting populated with vivid characters. Each scene is set in a specific location, the journey through them acting as a set of subheadings for the information stored. Sounds like a memory palace, right?

In the very early days of writing in ancient Greece, the skill was restricted to the elite. Ancient Greek and Roman societies gradually moved from being primarily oral to being primarily literate, aided by the introduction of papyrus from Egypt. The Greek gods—the heroes of Homer and Hesiod and the traditional legends of old—were no longer needed to impart practical knowledge in a highly memorable form; that learning could just be written down. The gods instead became a source for theatre, literary works and philosophical discussions.

The invention of the art of memory is most often attributed to the oral poet Simonides of Ceos, who was born around 556 BCE, but I think it's more accurate to say he was just the first to write it down, given there is no doubt that non-literate cultures had been using the technique around the world for far longer. Oral poets used to be hired as dinner entertainers; Simonides was the guest poet for a dinner held by Scopas, a nobleman of Thessaly. Various reasons are given for why Simonides left the banquet, but the important thing is that he left the room. While he was outside, the roof collapsed, killing Scopas and all his guests, and crushing their bodies beyond recognition. But the poet was able to recall where each of the guests had been seated for dinner, so the bodies could be claimed by their loved ones for burial.

Simonides realised that it was through the physical locations of the guests that he was able to identify them and that this orderly arrangement was essential for reliable memory. From this insight, it is said, he developed the method of loci.

Not long after Simonides' time, one of the big stars of the classical world took to the philosophical stage. The Greek philosopher Socrates was born in 469 BCE and didn't leave us anything written. It is possible that he was illiterate. We know his words only through his pupil Plato, who was born about 40 years later. In *Phaedrus*, written in about 370 BCE, Plato presents Socrates quoting a conversation between the Egyptian credited with the invention of writing, Thoth, and the God Thamus, the king of Egypt. Thoth said that the invention of letters would make the Egyptians wiser and improve their memories. Thamus disagreed:

> For this invention will produce forgetfulness in the minds of those who learn to use it, because they will not practise their memory. Their trust in writing, produced by external characters which are no part of themselves, will discourage the use of their own memory within them. You have invented an elixir not of memory, but of reminding; and you offer your pupils the appearance of wisdom, not true wisdom, for they will read many things without instruction and will therefore seem to know many things, when they are for the most part ignorant and hard to get along with, since they are not wise, but only appear wise.[3]

In turn, Plato's famous student Aristotle also wrote about memory training. Like the author of the *Rhetorica ad Herennium*, he suggested marking each fifth location in some special, memorable way. His students could then skip to each of these locations to get an overview of the structure of their philosophical argument, in order to analyse it.

I am very happy to follow Aristotle's advice. When setting up my memory palaces, I always set every fifth position first and then fill in the four gaps between. I find it really helps in establishing the palaces in my memory. I then say every fifth location when mentally travelling the whole journey. That grounds it for me and I return to the start to recall the entirety, or jump to the portion of interest at the time. There is a pattern and rhythm to the structure that becomes even more pronounced every time I name the locations aloud. Or sing them.

A highly trained memory was greatly admired in the classical Greco-Roman era. Not only was it useful in politics and speech making, it was also a terrific way to show off. These guys were the celebrities of their time. The Roman orator Seneca the Elder had 200 of his students recite a line of poetry of their choosing, one by one, each line unrelated to the last. Seneca recalled all 200 in order, and then proceeded to repeat them in reverse order. It is also claimed that Seneca could repeat 2000 names, having been told them only once. Pliny the Elder described numerous feats of memory, including Mithridates of Pontus, who knew the 22 languages spoken in his domains, and Charmides, who could recite the contents of all the books in the library, or so it is told. Then there was Simplicius of Cicilia, who could recite all of Virgil forwards and backwards, according to Saint Augustine. It doesn't record whether that was every letter backwards, each word in the sentences in reverse, or the verses backwards. But apparently he could do something really impressive with Virgil.

But it was Saint Augustine who, when he was mere mortal Augustine of Hippo, wrote a passage that resonates so strongly with me nearly 2000 years later, that he remains one of my most interesting 'ancestors'. In a brief passage in *Confessions*, Augustine describes retrieving images carefully placed in memory locations. Some images come back effortlessly with the mere thought of a location. Others require more effort. Those that are not well enough placed, or have little relevance, can be lost altogether. Augustine describes exactly the way experiencing a memory space feels. And he does it so beautifully.

And I come to the fields and spacious palaces of my memory, where are the treasures of innumerable images, brought into it from things of all sorts perceived by the senses . . . When I enter there, I require what I will to be brought forth, and something instantly comes; others must be longer sought after, which are fetched, as it were, out of some inner receptacle; others rush out in troops, and while one thing is desired and required, they start forth, as who should say, 'Is it perchance I?' These I drive away with the hand of my heart, from the face of my remembrance; until what I wish for be unveiled, and appear in sight, out of its secret place. Other things come up readily, in unbroken order, as they are called for; those in front making way for the following; and as they make way, they are hidden from sight, ready to come when I will. All which takes place when I repeat a thing by heart.[4]

Why not use both artistic *and* written devices? That's what they did in the early days of writing, when books were far more expensive than the one you are reading now. They were also far prettier.

CHAPTER 7

Lessons from the Middle Ages

The big lesson of this chapter is: don't make nice neat notes. Decorate and doodle all over them.

Logic might suggest that there is less to learn about memory methods from history after the spread of literacy across Europe. Everyone could outsource their memory to writing, after all. So I was very surprised when I discovered that there was actually *increased* interest in memory training during the Middle Ages. It was the *type* of information being memorised that changed significantly from classical times. No longer did the population learn utilitarian information from the orators. Medieval trade specialists and their guilds took control of most of the practical knowledge, much of which was kept secret, only available to the guild members.

Medieval folk filled their architectural memory palaces

with stories from the Bible. Then they made their ornate manuscripts extraordinarily memorable.

I adore medieval manuscripts. I think they are some of the most beautiful artworks ever produced. But self-indulgence is not the only reason I have included a chapter on the Middle Ages. There are valuable lessons for us today from medieval memory methods designed to serve alongside the written word.

Many medieval manuscripts are digitised and available online, such as the amazing collection on the website of the British Library. Be warned though, venture there and you may be lost for hours. One of my favourite examples can be seen in Plate 21, a page of the Gorleston Psalter.

The art changes purpose

The Middle Ages are generally considered to have begun around 400 CE and ended a thousand or so years later, when new thinking and ideas led to the Renaissance. The population slowly became more literate and books gradually became more available. But the memory arts were not lost. They just changed shape.

As in classical times, memory training involved associating information with emotionally striking images in a set of ordered physical locations. The rhetoric and rational arguments of the classical orators morphed into the pious sermons and meditation of a society dominated by the Christian church.

As the Western Roman empire collapsed at the start of the fifth century, the safest places became the monasteries, where education focused on the relatively new Christian creed. People with highly trained memories were regarded with awe, not only as a sign of intellect as it had been deemed in classical times, but also as a mark of superior moral character. That is why many descriptions of the lives of saints make reference to their astounding memory. Saint Francis of Assisi was apparently a great memoriser. Saint Augustine had a prodigious memory. Saint Anthony of Padua (1195–1231) is said to have memorised the entire Bible. Saint Thomas Aquinas (1225–1274) mixed in a highly literate group, yet his colleagues' greatest praise was for his memory.

Imagine that you are a student in the Early Middle Ages. You are expected to memorise huge portions of the Bible, word perfect, as well as important speeches of leading thinkers from ancient times, current speeches, laws, history, the natural history of animals and plants and a lot more. The knowledge is contained in books. There is only one small hitch. The only books are a few precious handwritten manuscripts. These are written either on papyrus or on parchment made from animal skins. Both are very expensive materials. There is no paper yet. You can use a wax tablet, but you have to constantly erase it to use again, so you can't keep your notes for revision. How can you remember all that you are expected to know?

You have to memorise it all.

Many medieval writers described the brain with metaphors that suggest memory palaces. Plato's ideas were still prominent. He had written of memory being like a pigeon coop, with pigeons representing bits of knowledge that are stored when we are able to shut them in our mental coops. Other metaphors for memory included bees bringing knowledge-nectar to beehives, coins being deposited in divided money pouches, treasures being placed in a chest and books being deposited on bookcases. Memory was seen as a structured set of locations into which bits of knowledge were stored. Coins, jewels, bees and pigeons and other memory tropes were frequently drawn in the margins of early medieval books. These images are usually unrelated to the content of the actual text but serve to make the page unique and memorable.

So recollection was seen as drawing together knowledge from different storage cells. It was considered an act of composing—one of creativity—that could not be performed unless the fundamental texts had been thoroughly memorised in advance.

I often worry that we sacrificed some of this creative power when we started depending on the written page to store all our knowledge and memories. We only know the information in front of us. Have we lost the big picture available to earlier thinkers, who had the grounding of any subject available to them in their memories? Are we now making it worse by even more entirely depending on digital memories rather than our own? I find my thinking so much more expansive now that I have all sorts of structures in memory to build on. Can't we optimise our thinking by making the best use of all three: memory, writing and computer technology? Are students who can do that going to be the leaders in an age where the ability to create new knowledge becomes more and more critical?

But I digress. Back to the Middle Ages. The preachers who toured the country delivering greatly entertaining sermons were dependent on the same techniques as the orators of the classical world: they broke down their speeches into small segments and associated each segment with some portion of architecture of an orderly and defined space. Monasteries were particular favourites as memory palaces.

Are the wondrous, ornate, grotesque, vivid sculptures and other artworks that adorned the monasteries as much about making everything memorable as they were about what appealed to the taste of the times? And those bizarre images flowed to the manuscripts for which the Middle Ages are so famous.

Medieval lesson 1: Make every part of your page look different

Highly decorated text with stylised artworks became memory spaces for religious knowledge in both the East and West of the Roman empire.

The beautifully inscribed handwritten words were enmeshed in flourishes. It was common to have each chapter start with a coloured initial, alternating between red and blue. There were various standardised forms for the commonest letters so the same visual impact of the letter would not be repeated on a single leaf of parchment. Distinctive ornamentations were added around the text.

Both monks and nuns illuminated the manuscripts in the Early Middle Ages. Very little of these manuscripts was new: they copied the same religious texts over and over. Most commonly, they copied the individual books of the Old Testament, the first four books of the New Testament and the 150 psalms into collections known as Psalters. Monks were expected to memorise, as an absolute minimum, every single psalm. That alone took somewhere between six months and three years.

Medieval lesson 2: Add emotion to everything

The texts were adorned with gold leaf, enriching images of grotesque and violent acts, fanciful beasts, strange figures, gross ugliness and extraordinary beauty. But most important of all, the pages of the text had to stir the emotions to make the written word unforgettable.

The large books were set on lecterns in churches and monasteries, giving melodramatic touches to theatrical church services as the glorious books were opened and the magnificent pages turned.

Medieval lesson 3: Lay your information out in grids

By 331 CE, the Roman historian Eusebius of Caesarea divided the gospels into around 1160 sections. He then created lists of chapter numbers, known as 'Eusebian canons', so that monks could match up parallel passages in each of the four accounts of Christ's life: the gospels of Matthew, Mark, Luke and John. The canons indicated areas of agreement and difference between the gospel accounts.

The canon tables were designed to be memorised, and their basic design remained in use for nearly a millennium. They were usually presented as shown in Plate 22. The elaborately decorated lists of numbers were written between illustrations of columns with arches above, reflecting the ancient memory advice to use inter-columnar spaces as locations for memory images. The vertical spaces between the columns were then divided by horizontal lines into small rectangular spaces, each holding no more than five items, the maximum number suggested for retaining in memory for a single location.

Laying out the narrative in a grid of images makes it more memorable. Your brain will remember where a given rectangle in the grid lies in the space and hence recall the information. This is reminiscent of some of the mandala layouts mentioned in the previous chapter. One of the oldest European books is a copy of a sixth-century manuscript of the gospels believed to have belonged to Saint Augustine. Many of the stories are painted in grids, some of the most famous examples being three cells by four cells, as in Plate 23. The images are not only unique but positioned in a unique location on the page.

Hundreds of years later, the same advice was still being given. One of the most influential writers on how to memorise was theologian and teacher Hugh of Saint-Victor. (We first met him and his virtual memory palace based on Noah's Ark in Chapter 2.) Born in what is now Germany, in 1096, Hugh became an Augustinian canon before travelling to Paris, where he settled at the abbey in Saint-Victor, teaching in the school there until his death in 1141.

Hugh recommended using grids of cells to memorise large swathes of text. For example, for the 150 psalms, he recommended that the beginning phrase of the first verse be placed in a cell. The cells were placed in a line of 150 locations. For each psalm, he then imagined another set of numbered cells, one for each verse. These are the numbers that probably became the current numbers for biblical verses. They were not added into manuscripts until the middle of the sixteenth century, 400 years after the numbering was first documented.

Of course, you could get really enthusiastic and design stained-glass windows for your home based on the narratives of knowledge you want to share. Many church windows were laid out in grid structures to make the narrative easy to follow for the illiterate congregation. Medieval churches boasted glorious colourful images in sequences of stained glass, each telling a small part of the story. Staring at those superb windows week after week ensured that the stories of the Bible were well entrenched in medieval minds.

Medieval lesson 4: Give character to abstract concepts

Martianus Capella's *De septem disciplinis* ('*On the seven disciplines*') was written between 410 and 429 CE, so strictly speaking it dates from the Dark Ages, but Capella's work was used as the standard for academic learning for 700 years, well into the Middle Ages. That's the equivalent of teachers today using teaching methods written before the invention of the printing press. He must have said something that people found worked well.

Capella's work introduces its reader to the seven liberal arts: Grammar, Dialectic, Rhetoric, Geometry, Arithmetic, Astronomy and Music. These are pretty abstract themes, so he employed the really ancient method (even in his day) of telling a vivid story in which the seven liberal arts are characters: at the wedding of the Roman god Mercury and Philologia (a name meaning the love of learning), seven maids were given to the bride. The seven maids are the seven arts in the form of people. Grammar was a fearsome old woman seeking grammatical errors to remove from children's work with her knife and file. Rhetoric was tall, beautiful and richly dressed in cloth decorated by figures of speech and carrying weapons with which to wound those who would argue with her. The personifications of Architecture and Medicine were also present, as they were taught in schools, but considered unworthy of mention in the company of the deities.

Whenever you need to learn an abstract theme, give it a character. Electron is a really negative guy. Commas are commoners. Fences are fencers slashing with their swords. Rectangle is a woman who hates anything oblong and so smashes the angles until it is some other shape. When you then weave stories around the character, the associated facts become memorable.

Medieval lesson 5: Break it down into small portions

During the High Middle Ages, after around 1000 CE, the population expanded rapidly. Books became more widely available and cathedral schools became more common, moving the hub of education out of monasteries. The first universities were established in Italy, France, Spain and England, starting with the University of Bologna which was founded in 1088. The memory arts were part of the standard curriculum. All academics and students were expected to justify their arguments from a store of memorised facts, a skill that became particularly important for lawyers.

The secret to memorising anything is to break the information down into memorable portions; just focus on a snippet at a time. This is demonstrated in the structure of the Bible. The Bible was first divided into memorable, short divisions in the Middle Ages. They were continually adjusted: Stephen Langton, who served as the Archbishop of Canterbury from 1207 to 1228, spent 30 years perfecting the chapter divisions, testing them repeatedly in his classroom teaching. Further division in the sixteenth century led to the current arrangement of book, chapter and verse.

The efficacy of short sentences on a memorable page resonates with my experience as a teacher. I have found that students who read an entire paragraph of information quickly will often claim they didn't understand it, but if they read it phrase by phrase, stopping at each comma or full stop to ensure they understand, the entire paragraph becomes meaningful. With short sentences, you are forced to engage with each element of the information and not try to grasp the whole in a single befuddling quest.

A page of neat and tidy typed text in long paragraphs is the least memorable format known. You need to reduce it into small segments, each made memorable by flourishes and fancy layouts. Add colour and doodles. Highlight. Enclose with clouds. Write the whole portion backwards. Do anything to make each logical entity, each verse, distinct.

Medieval lesson 6: Separate those short portions on the page

Hugh of Saint-Victor—he of the Noah's Ark memory palace and feats of psalm recitation—had a lot to say about memorising. He instructed that short sentences should be memorised exactly as they appear on the manuscript page. Students should always learn from the same manuscript because the memorisation was so heavily based on the visual image of the text. They should take special note of the colour, shape, position and placement of the letters, their location on the page and the ornamentation. He believed there was no point in education if what was learned could not be memorised.

Contemporary educators might disagree with that final sentiment, but his position-on-the-page bit is consistent with the comments of musicians in Chapter 8 about the importance of working from their own familiar score when performing.

So many people have told me how the position on the page in a book, newspaper or magazine comes back when trying to recall some fact they've read or an image they've noticed. Why not take advantage of this mental trait and place each detail at a clearly defined part of the written space?

Medieval lesson 7: As always, use memory palaces

You may have noticed that the early medieval lessons were all about grids and alphabets, animals and the zodiac and glorious decorations around the written words. The use of architecture and the physical landscape to store memory images seemed to have been put aside in medieval times. But two major figures of the High Middle Ages, Albertus Magnus (c. 1200–1280) and his pupil Thomas Aquinas (1225–1274) brought the method of loci back into fashion.

Albertus of Cologne, as he was known during his lifetime, founded Germany's oldest university at Cologne. He was an influential academic in fields including science, philosophy and theology. He was also a highly respected diplomat. Albertus studied the writings of Aristotle and the contemporary teachings of Muslim academics (particularly Averroës, one of my 'ancestors'). He argued that the anonymous work from 85 BCE, the *Rhetorica ad Herennium*, provided the best memory system of all, and so memory palaces came back into the training repertoire. The anonymous author of *Ad Herennium* continued to be read into the Middle Ages, 1300 years after he had penned his school textbook. We authors dream of such an impact.

During the Middle Ages the *Ad Herennium* was ascribed to 'Tullius', the name by which Tullius Cicero was known. This seemed to have happened because the *Ad Herennium* had been tacked on the end of a publication of Cicero's *De invention*, which was also still in use.

In the pious Middle Ages, violent, lewd and fanciful images were deemed highly inappropriate. I am delighted to report that Albertus justified their use because, ironically, they were so effective for memorising moral philosophies. His pupil Thomas Aquinas was to become one of the most revered saints of the Catholic Church. In her seminal work, *The Art of Memory*, Frances Yates wrote: 'If Simonides was the inventor of the art of memory and "Tullius" its teacher, Thomas Aquinas became something like its patron saint.'[1]

Born in 1225, Thomas Aquinas had a phenomenal memory even as a schoolboy in Naples. With the advantage of memory training with Albertus, he was reputed to be able to perfectly retain everything he read. He wrote that, like Aristotle, no human thinking could take place without images.

Medieval lesson 8: Meditate upon your memory palaces

Even though Thomas lived in a time when books were more readily available than during the Early Middle Ages, he still memorised all his reading to enable him to draw information in any combination he wanted. He used memory palaces. He emphasised the need to meditate on each location frequently, writing that:

> Meditation is nothing other than considering things many times as an image of things previously apprehended and not only in themselves, which mode of preserving pertains to the

formality of memory. It is clear, too, that by the frequent act of remembering the habit of memorable objects is strengthened, as also any habit (is strengthened) through similar acts; and a multiplication of the cause fortifies the effect.[2]

That's the big lesson from Thomas Aquinas: meditate. Go over your journeys and palaces, your memory boards and songs, but do it gently and slowly. I often do this when going to sleep. If you can slow yourself down to think about only one or two locations, this form of meditation is incredibly relaxing. The more often you contemplate your memory locations, the more they will become an integral part of your thinking and knowing.

Throughout the High Middle Ages, the memory section of the *Ad Herennium* was still widely read. The Dominican friars, in particular, recommended the memory arts for learning lengthy lists of virtues and vices and so avoiding eternal hellfire. Memory treatises were translated from Latin and circulated in the vernacular languages, especially Italian and French.

Not long after Thomas Aquinas's time lived the first genuine bibliophile: theologian, teacher and writer Richard de Bury (1287–1345). He created most of his library by memorising books. If de Bury could not buy a book because too few copies existed, he would memorise the entire text in someone else's collection, return home and have his secretaries write it out from his dictation. He also had a network of clerics who would travel all over the Christian world, memorise books and recite them back to him. In this way, books that resided in one city of Europe could travel, via a trained memory, to any other location. Copyists and secretaries were a growing profession as they beavered away replicating books. The monks and monasteries were no longer in control.

De Bury is also remembered for his *Philobiblon*, considered the first text on librarianship, but I think using riches to employ people to travel and memorise books is a far more intriguing call to fame.

Medieval lesson 9: Decorate your walls, but do it systematically

Andrea da Firenze painted a wall of the Spanish Chapel in the church of Santa Maria Novella in Florence sometime around 1366. The fresco is *The Triumph of St Thomas Aquinas*, also known as his *Wisdom*, shown in Plate 24. It reflects the principles dictated at the time for making images memorable.

Thomas is seated in the middle surrounded by three winged figures representing the three theological virtues—Faith, Charity and Hope—along with the four cardinal virtues—Prudence, Justice, Fortitude and Temperance. On his right and left sit the saints and patriarchs. Beneath Thomas's feet are the heretics, who have been crushed by his wisdom.

On the lower levels are fourteen female figures, the bodily representations of the abstract seven liberal arts—Grammar, Rhetoric, Dialectic, Music, Astronomy, Geometry and Arithmetic—and Civil Law, Canonical Law, Philosophy, the Holy Scriptures, Theology, Contemplation and Preaching. Cicero sits in front of Rhetoric, while Saint Augustine sits in front of Preaching. There is some doubt about who's who among the rest but it is generally thought that Aristotle is the figure in front of Dialectic and Philosophy. Other figures represent Thomas's vast knowledge of lofty themes.

We are back to the methods described in Chapter 1, which inspired my visual alphabet. It was Thomas Bradwardine who described constructed scenes with each character linked to those nearby. He noted that the associated knowledge can be recited from any starting point, forwards or backwards.

Medieval lesson 10: Leave room on your notes for additions

Please do not follow the example of the medieval scholars when using your library's most precious and stunning holdings: they scribbled in the margins of the most awe-inspiring artworks of all time. They actually wrote on the illuminated manuscripts, as shown in Plate 25.

During the twelfth century, readers started writing in the margins of the manuscripts to add their own commentaries, known as 'glosses'. The books were designed specifically for memorising and left plenty of space for readers to have their say. The glosses weren't just random notes but additions from sources the scholars considered had authority and would add more to the text already there. The source text was written on alternate lines in the centre of the page with the glosses in smaller writing between and around them.

The complexity of the illuminations, drawings and pictures in medieval manuscripts was deliberate, encouraging the reader to ponder them at length and repeatedly. The goal was to commit them reliably to memory.

Manuscripts did not have indexes, so scholars memorised a variety of signs and abbreviations in the margins to indicate a link to a particular topic. They used all sorts of symbols, anything that had a specific order, including foreign alphabets, zodiac signs and mathematical figures. Some Roman legal specialists, known as notaries, apparently knew over a thousand signs.

Medieval lesson 11: Add playful little drawings

Quirky illustrations, known as drolleries, were also frequently added around the text to make each page even more memorable. Unlike glosses, there doesn't seem to have been any relationship between the drolleries and the text. The drolleries often depicted foxes and roosters, cats and mice, dogs and hunters and savage boars. Some manuscripts had entire narratives in pictures along the bottom of the text, or even between the columns. These could be stories of animals, such as Aesop's Fables or those of Renart the fox. There might be comical images of animals, such as monkeys imitating human behaviour. An example can be seen in Plate 26, created around 1340. They were usually humorous, violent, ugly or titillating— anything which would invoke an emotional response and aid memory. A plain page of text surrounded by clean space was considered very poor form.

I strongly recommend visiting the British Library website, where the Smithfield Decretals have been digitised in all their glory. Warning: do not attempt this when you are short on time. Hours may disappear as you become engrossed in one of the most stunningly beautiful books ever produced.

The Decretals are digests of canon law and contain the most wonderful images, unimaginable in a legal text of today. Students were expected to memorise the Decretals in the order given and with extreme accuracy. The bottom margins were adorned with delightful stories told entirely through pictures, resembling the frames of a cartoon. Killer-rabbits, for example, track a hunter, take him to court and hang him. Another sequence tells a scandalous story of the seduction of a miller's wife by a friar who then murders her husband, while other drolleries vividly depict grotesque incidents from saints' lives.

If you are working entirely from neatly typed notes, you are making your life really difficult. They are so dull to look at. Handwritten notes will serve your purpose far better. But even if you use typed notes, print them out with space around them. Lots of space. For any notes, typed or handwritten, leave wide margins and gaps between paragraphs. When revising, doodle. You can add images and comments around written text, mathematical proofs, scientific analysis or any other form of knowledge.

Anything you draw will make you slow down and engage with the actual information. Add little images that make puns, grotesque stories and anything else that adds character to abstract ideas. If learning about a country, add a rabbit waving their flag. When studying the *Mona Lisa*, sketch her moaning. Surround your Pythagorean triangles with a python trampling any triangles that do not have a right angle. Make every page unique. It doesn't have to be flash, just memorable. The page will become your memory palace.

And it is fun!

My medieval manuscript on musical instruments
I love studying, so I still spend hours learning new topics and I use the medieval lessons above for all my notes. To begin with I just added scribbles and scrawls. But then I got a bit carried away. You don't actually need to go so far as to learn gothic

script and illuminate your capital letters in gold in order to adapt the memory techniques from the Middle Ages.

I realised that almost any topic would do for my illuminated medieval manuscript. The official story is that I chose the history of musical instruments because music is such a critical theme in human history and science. The real reason, I feel forced to confess, is that I think musical instruments are absolutely beautiful and I wanted an excuse to draw them.

Memory treatises of the Renaissance

During the Renaissance, the spread of literacy and availability of books was greatly increased with the introduction of the printing press. I assumed that these radical changes would destroy the need for memory systems, but there was in fact a sudden surge of short books about mnemonics, known as memory treatises, or *ars memoria*. Jacobus Publicius published the first printed memory treatise in 1482, titled *Oratoriae artis epitome*. Emphasising use of places and images, it was to be the first of many printed books on the memory arts. They taught the familiar method of dividing rooms into five locations, which is exactly what I do when teaching workshops. It is the best starting point for those new to memory palaces. Use your own home.

The most popular Renaissance treatises included those of Johannes Romberch and Peter of Ravenna (the ones with the visual alphabets and drawings of the abbey I talked about in chapters 1 and 2). At long last, readers were freed from the medieval practice of endless references to vices and virtues, heaven and hell. Peter gave practical advice and examples that appealed to a lay audience. It is no wonder that his books sold well.

Having started in the Renaissance in Chapter 1, we've now returned there. Now we'll turn to our own time and explore the implications for the young and old of maintaining a healthy memory right into old age.

CHAPTER 8

Learning in school and throughout life

Memorisation certainly isn't the main goal of education, but it is a critical component of a suite of invaluable skills. This chapter is about committing basic knowledge to memory in schools, university and throughout life. I believe that you need a fundamental knowledge base on which all the glorious creativity, analysis, evaluation and critical thinking can build.

I have often been asked why we should bother memorising when we can just look things up. It seems there are people who are perfectly happy to get all their knowledge from the internet, which leaves them seriously vulnerable both emotionally and financially to those peddling false news, fake cures and fantastic lies. If nothing more, memory is essential in so many careers. I checked online for lists of the most respected jobs. How much memory work is involved in the professions below? As you read the list, ask yourself the following questions: Do you need to memorise course material to qualify for these professions? Would you feel comfortable relying on any of these professionals if they turned to Google every time you asked a question? Which of these professions do you imagine will disappear or be taken over by robots in your lifetime?

The internet consistently offers the following professions as the most respected: physician, nurse, lawyer, computer systems analyst, teacher, physicist, astronomer, chemist, pharmacist, architect, biological scientist, ambulance officer, engineer, university lecturer, academic, judge, chief executive, public administrator, geologist, psychologist, manager and clergy.

I hunted list after list for 'author', but it wasn't there. I suppose many genres of writing could manage without a permanent knowledge base. But how far could you go in your career as an actor or singer or performer of any kind without being able to memorise? If you are required to speak in public in any context, your audience will be unimpressed if you simply read from a printed sheet.

Memorisation is not a goal in itself, except if you want to star in quizzes or enter memory competitions. Memorisation is a way to enhance what we already do in education and in life.

I have spent four decades in classrooms, mostly teaching senior high school mathematics and science, but also subjects right across the curriculum, from primary schools through to university. The pendulum has not swung entirely away from teaching knowledge. Good teachers blend the more complex thinking skills with the factual content of their curriculum.

I need to be quite clear on one point: memorisation is not rote learning. There are really important differences. Rote learning is defined as learning by repetition without any regard for understanding. If you learn Tom Lehrer's song 'The Elements' then you will know the names of all the chemical elements by rote. He sings them in an order that aids the rhythm of the song (which I love) but that sequence tells you nothing about the chemistry. If you store the elements in a memory palace based on the periodic table, as described below, then you are creating a memory device that will allow you to build knowledge of the chemistry into multiple layers of complexity.

In the last chapter I mentioned 'modality shifts'. I believe exercises based on this theme are hugely effective. You take information presented in one mode—say, writing—and shift it to another mode, such as art, story or song, or embed it in a memory palace. Asking students to shift the information from one mode to another forces them to engage with it. A significant part of memory is concentrating in the first place.

Once students get good at making up stories, they can make anything, no matter how dull, into something interesting. The goal is to let their imaginations go wild.

The following are just some examples describing how various memory methods from earlier chapters can be used in schools, universities and, of course, for lifelong learning. The table of memory methods in Appendix A will help you evaluate which method suits your particular subject and the genre of information you are trying to memorise.

Permanent memory palaces for all students

Given that memory palaces are by far the most effective memory devices known, why don't we use them in schools and universities?

In fact, why don't we use our schools and universities *as* memory palaces? School corridors are busy briefly between classes but empty during them. Nearly every school I've been in has had vast corridors of empty walls and classrooms neatly numbered, one, two, three . . . What if the classrooms were labelled according to something just as easy to navigate but also useful? Maybe, O CE, 500 CE, 1000 CE, 1500 CE?

At 1066 CE, there could be an image of the Norman conquest of England. It could be a serious poster, or it could be an image by students having fun with puns and jokes and even puzzles. Anything which has the Norman Conquest just past the doorway labelled 1000 CE is going to put that event at the right time in history without having to memorise the exact date.

Classrooms in another corridor could be labelled 1800, 1850, 1900, 1950, to allow detail for more recent events. Students from any subject could attach significant images at the appropriate stages along the corridor. The walls could be cleared regularly, giving an ever-changing experience, while the key features, the dates, remain unchanged. New items will attract attention, and those that have become a little worn can be removed.

Or the doors could be labelled A, B, C and used as a dictionary for any language that uses the Roman script. The wall at A could be a pinboard for images relating to words starting with that letter. Artworks could be created in art or communications classes, the skills of those teachers adding greatly to the display. It doesn't need to be simply drawings—there could be sculptures or any other art form. You want to recall the French word for 'apple'? You remember that it is down the corridor on the second floor, with words that start with a 'p'. *Pomme.*

Similarly, school grounds are the perfect venue for a permanent memory palace. A post at each location will enable students to refresh their knowledge of their palace whenever they like. Even when they have left the institution, that palace will be firmly in memory.

At a rural primary school, art teacher Paul Allen and I worked with students to create a History Trail in the school grounds. We started at 1800 CE, just before the first Europeans arrived in the area. Paul made posts to stand at every ten years until the present. We associated events with locations, sometimes marked by sculptures made by the students, at other times by existing features. We linked significant local, national and global events. We adults put in the world wars and the founding of the school. Students added the first flushing toilet in Australia and the start of Australian rules football.

We walked the entire loop with the students, who got very excited as we neared the end and they were finally born. One student was particularly interested in Indigenous issues. At the start of the trial, he tied his hands together with the arrival of the first Europeans in the area, representing the Aboriginal people who were not recognised in the Constitution. To Australia's shame, they were not even counted in the population. As we walked, he kept commenting that he was not yet a person. He mentioned that he was allowed to serve in the armed forces as we passed each of the world wars, but was still not a person. We passed the year when I was born (they sang happy birthday to me) and the birthdates of the other staff members, grandparents and some of the parents. We had travelled most of the way around the school grounds when we arrived at 1967. This was when Australian Aboriginal people were finally counted in the Australian census. The student untied his hands, announcing that at last he existed. All the other students cheered. His sad walk had far more impact than any lesson reciting dry dates.

At the end of term, the parents and grandparents were invited to walk the History Trail with their children and grandchildren. The students were able to draw their families' attention to events that had interested them while the adults added when they had been born and stories of their lives. History became enmeshed with their personal saga. The History Trail can be constantly enriched with scientific discoveries, political events and personal passions. I will always associate the oldest flushing toilet in Australia with the big blue bin at 1900.

Using the same memory palace for science and fine arts

A single memory palace can serve a multitude of purposes; there is absolutely no need to have a separate memory palace for every topic. At one secondary school, we set a palace around the outside of the school buildings; each door was the fifth location and we found four features in between: windows, taps, garden beds. We used that same memory palace for the periodic table in junior science and for the structure of the senior Visual Arts curriculum.

The Periodic Table

Any chemistry student who has to look up an element every time they need to use it in a chemical reaction or understand its bonding is going to be painfully slow. For many elements, their position in the periodic table will tell you most of what you need to know. For others, you need to know the boiling and freezing points, isotopes, decay series, natural occurrences and major uses.

It was fairly easy to associate the elements with the locations. Hydrogen was a bomb and we blew up (in imagination only, I promise) a plaque declaring a new wing open. Helium balloons flew from the staffroom window, while lithium became lithe young bodies doing sport. These were teenagers after all.

It was fairly quick to recall the atomic number of any element from the palace structure. But there's a lot more to the periodic table than atomic numbers. Each column represents a Group, which gives significant information about the chemical behaviour of the elements.

When we reached an element in Group 18, the noble gases, we bowed out of respect for their nobility, which made them highly memorable. Neon—element 10—is a noble gas: every student bowed at location 10 and then put their 'knee on' the door, some a tad more violently than others. That action, performed only once, will fix neon at location 10 forever.

Knowing the location of the noble gases makes it easy to know the Group of any element in 17 or 16 by going backwards, or Groups 1, 2, 3 or 4 by going forwards from the nearest Noble, as shown in Figure 8.1.

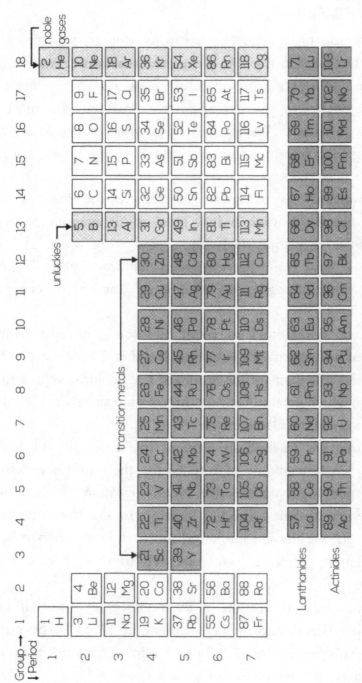

FIGURE 8.1 The periodic table with the Noble Gases, Unluckies and Transitionals shown.

At a Group 13 element (the Unluckies), we crossed our index fingers to scare off the bad luck of 13. That helps you to place Groups 14 and 15 easily. It is really quick to scan the memory palace for the Nobles and Unluckies to know where any element sits in the table.

The transition metals first appear in the fourth Group. We folded our arms through that section, representing a transition, going across.

At the first school, we didn't get as far as element 57, lanthanum. At another school, the science enthusiasts wanted to do the entire table. From location 57 on, there are two sets of elements that don't fit in the table and are always put in their own section. They still appear in the memory palace in order so the atomic numbers are never lost. But there are no Nobles or Unluckies in their ranks.

The first set is the lanthanides. We sang those elements, starting with 'La, La, La lanthanum'. If an element is sung, it is a lanthanide. For the second, the actinides, the students created a play to 'act' in that section of the palace. One group of students created a play in seven minutes about Thor (thorium) getting uptight about a prostitute (pro-tactinium) carrying a uranium bomb to blow up Neptune but his dog, Pluto, then . . . and so on. There's a lot of radioactivity in the actinides so it was a fairly explosive play.

The elements and principles for visual arts

The same memory palace was then used by the visual arts students to memorise their curriculum, which is based on the art elements (line, colour, tone, texture, shape, form, sound, light and time) and principles (balance, contrast, emphasis/focal point, movement, proportion, repetition, rhythm, scale, space, unity and variety). These ideas are integral to the students' vocabulary and need to be considered whenever they are analysing or creating artworks.

At the first door, the fifth location, we needed to add 'shape'. Their teacher pointed out the geometric shape of the door contrasting with the organic shape of their own bodies reflected in the glass. Within half an hour, the students had a similarly concrete image for each element and principle. They could quickly scan the palace when thinking about their analytical essays or considering their own creative work.

Students can use the same palace for multiple topics, making each location more and more vibrant. They won't get the images confused.

The memory palace is a grounding for a theme and claims to be nothing more. Teachers and students can then build a much more complex structure on the firm foundation, bringing in all the higher levels of learning and thinking.

One really exciting glimpse I had into the potential of memory palaces and schools was when I worked with students with learning difficulties. Many special needs teachers have suggested that memory palaces may enable dyslexic students to learn more easily. Information is usually presented as writing, which is the crux of their disability. Being familiar with the information through other memory methods gives these students a guide to reading the words. Although this seemed to be true for the few dyslexic students I worked with, only rigorous research would justify this claim.

I saw many disengaged students become very excited by memory palaces. They were able to use their personal passions to create weird links between the images and stories. Students commonly used cartoon characters, pets (alive and long passed), toys and, far more than I would have ever predicted, imaginary friends.

Using song, stories and the wonderful rapscallions

The vocabulary for every discipline can be grounded in song. At a primary school in rural Victoria, the students were all learning about force; it is the basis of the state-defined curriculum for a few weeks right across the year levels. A week after their initial lesson, I asked every student the same two questions.

'Do you remember doing force in science?'

They all answered, 'Yes.'

'What is a force?'

Only three out of the 70 told me it was a push or a pull. The rest talked about the way a toy truck zoomed across the room and the balloons whooshed through the air or about parents, teachers or friends who force you into doing stuff. A large proportion said, 'May the force be with you!' It's a terrific slogan but really bad physics.

Then the music teacher Joseph Bromley wrote a short song about force, with actions to the tune of 'The Imperial March' from *Star Wars*. It was performed once in music class and again at assembly. The song had not been mentioned again for over a week when I asked each of the students exactly the same questions and every one of them told me that a force was a push or a pull and did the actions. Most sang and nearly all laughed.

Before the song, the teachers had been building on sand every time they used the term 'force'. A simple memory technique grounded the definition.

I will constantly return to my beloved rapscallions, intro-duced in Chapter 4. Adding characters is a fantastic technique for making any topic relevant to students. At one school, the students had written persuasive essays. The teacher placed a rapscallion in front of each student and they were told that the rapscallion disagreed with them totally. The students rewrote their texts to be even stronger when they had a character to argue with. They were then asked to write the rapscallion's point of view. Again, it was much easier for them to write a strong essay when they had to give voice to a physical character.

At another school, a teacher asked that I look at ways to help students who really struggled with spelling. The teacher particularly wanted to convey that the silent 'e' on the ends of words changes the sound. He described it as making a vowel 'say its name'. The vowel sound of 'a' becomes the sound of the name of the letter 'A': 'rat' becomes 'rate', and 'can' becomes 'cane'. For all our spelling work, Paul Allen, the art teacher, had made an illustrated alphabet and given characteristics to all the letters. The 'e' was energetic while the 'g' was gentle. To explain the way that 'e' makes the vowel say its name, the students illustrated a scenario demonstrating that when their rapscallion dumped the energetic 'e' on the end of the word, the vowel told their rapscallion its name rather than making its softer sound. The lesson took only about a quarter of an hour, but weeks later when I quizzed students they could all tell me what energetic 'e' did when you placed it on the end of a word.

Adding a concrete character to an abstract concept changes mundane information into highly memorable information.

For some reason, many educators frown on anthropomor-phism, but I love it. When teaching electric currents, I used to have my physics students dance around the room as electrons in a current. They were attacked by thugs (resistors), who tried to take away their jewels (joules, a lucky pun for me). The electron students had to return to the start of the circuit, having given up all their joules to the resistor students, or the battery would bash them up. (I called that assault-in-battery. They groaned at my jokes but didn't forget them.) They never confused the story with reality, but they also never forgot the physics. Looking back, that was a simplistic use of an incredibly effective memory system. If only I had known more memory techniques then.

There are so many opportunities to add creativity and imagination to the abstract concepts across the disciplines. Why do we keep them all so separate? I have been in many classrooms where a creative writing topic could very easily be adapted. For example, a creative writing 'story starter' showed an image of a door and asked students to write what was behind it. In science, they were learning about the planets. A quick melding of the topics, and the door became the entrance to a dinner party the sun was giving for the planets. The planets arrived in order. Jupiter was a stormy guy while Saturn was adorned with rings. Curriculum requirements in both subjects were satisfied.

Let's sing, dance and make musical memories

Music looms large when talking about memory and education, for three distinct reasons.

Firstly, I have addressed the use of song for memory often in this book because knowledge that is sung is remembered far better than unadorned prose. We sing the alphabet and other themes with very young students, and then music somehow moves to the periphery of the curriculum. Why? This does not happen in indigenous cultures. They just keep singing. I believe encoding information in music is such an incredibly effective tool that we should use it across the curriculum, throughout school, throughout university and throughout life. I want the whole world singing, even if it is sometimes a bit out of tune.

Secondly, people often ask how musicians use memory methods to help them remember the music, songs and performances in general. Can we take what they do with muscle memory and use it in other spheres of learning? I was convinced the answer was yes but first I had to make sure I understood what musicians meant when they talked about muscle memory.

And thirdly, what about learning music itself? The more I talked to musicians about learning music, the more I could see how their ideas could translate to any subject. We can sing and dance almost any topic. Why isn't music integrated more into all forms of learning? I'll let the musicians speak for themselves. You'll see lots of familiar themes from the previous chapters emerging in different guises.

Kirsten Boerema is the Music Coordinator at Castlemaine Secondary College and an experienced performer in her own right. She uses many memory methods in music classes. A simple mnemonic is all that is needed to remember the various scales associated with the notes C, D, E, F, G, A and B respectively: Ionian, Dorian, Phrygian, Lydian, Mixolydian, Aeolian, Laerian. Kirsten uses the sentence 'I Don't Play Lydian Modes After Lunch'.

For more complex ideas, Kirsten uses characters and emotions. She explained to me:

When teaching three C major key signatures, I give a character to Major, which is happy, and to Aeolian, which is a little bit sad. For the Harmonic Minor, the character adds a bit of bling because it has a chromatic note, a sharp seventh which doesn't quite fit. I draw a big earring on it. And then I add the Melodic Minor, which has a bipolar personality because it is different on the way up than it is on the way down. The emotions and the characters make the theory more memorable.

Kirsten also talked about the importance of the music on the physical page:

We had a not-musician music teacher. She was not successful as a musician in the real world. She didn't understand how the kids usually learn music because she's not a performer. She gave out sheet music that was a dreadful arrangement. The kids practised it. Four bars didn't work. She collected the music, changed it on a computer program and reprinted. She was very frustrated because the kids then couldn't play it. It was like them having to learn a new piece. She should have crossed out the four bars. They could leap, skip those bars. It would still be the familiar music in the same places on the page. They would have already added symbols like bowing markings and breathing markings but these were now lost. Maybe they had added a dot to symbolise a short note on the second bar of the third line. They know where that short note is because they drew it there. Now that note has moved.

This is reminiscent of medieval manuscript writers and the importance they placed on the image of the page for memorising the content. As described in Chapter 7, they would add drolleries and other little markings to make each page unique and unforgettable.

Many musicians I spoke to described the process of learning a piece for performance and annotating it to make the music their own. There are universal symbols used, such as a pair of glasses drawn on the music to warn them to be careful. Although sometimes printed on the music itself, musicians often add it themselves to know that this is the bit where they often mess up.

As opera singer Catherine Carby said:

It's like I can always see the score before my eyes and I'm just sight reading what I see. I also see the page turns and stuff I've written on the page and it totally throws me if I look at a different edition of the same score.

Fellow opera singer Jacqui Dark agreed.

Or worse, if it's a copied score and some versions of it have the pages around the 'wrong way'—pages that I have on the left side are now on the right side and vice versa—it does my head in. If I hear a page turn where there's not one, it throws me completely!

We got onto the topic of muscle memory. For Jacqui, 'what . . . singers call "muscle memory" is a kind of cross between two things':

In tricky pitchings, you can feel where the note 'sits' in your throat and cords, and just basically aim for that feeling rather than the actual note.

Secondly, you sing a phrase in so often (especially in a big coloratura section with millions of notes) that your brain becomes removed from the equation and your muscles go on autopilot to guide your voice through a phrase they know well. In fast phrases, you don't have time to think out notes, so you just have to repeat them until your cords do it automatically.

What about for instrumentalists? Kirsten described the general feeling as 'your fingers just know where to go—they are in autopilot'.

Practice is creating muscle memory. Your body remembers how to play it. With a new piece you just read the notes and play it. You're not going for beauty of tone or articulations like accents, staccato or legato. Neither are you going to be focusing on dynamic control like crescendo or decrescendo, because that takes great control. You need to learn the notes, the pictures. Is your finger in the right place? Is it in the tune?

Once you know how the pitches sound, you will repeat that as often as you need until you know how the piece should sound.

Bec Heitbaum is a freelance wind player.

I started on the flute. All I needed to know was where I put my fingers to be in tune. There's no vibration feedback from a flute like I get with the double reed instruments.

Now I play oboe. I can tell by the feel of the vibrations of the double reed if I am playing the right notes. I can feel if it is correct or if I've stuffed up. It's not an exact science and varies from instrument to instrument.

I had to hire a cor anglais for *Les Mis*. On that particular instrument, the F wasn't guaranteed to come out in the right register. If I was going for a low F, it was easy to over blow and take the F up an octave. However, I could feel the difference in vibrations. I went 'Oh bugger!' when this happened initially, but after a while, the different feel was committed to muscle memory. I paid attention to how it felt, not consciously, but as part of my subconscious memory. Because you can't hear yourself playing in the orchestra, I became sensitive to the vibrations as part of the muscle memory of playing that music on that instrument.

Bec emphasised that it's only once you've got the notes safely in place that you can add dynamics and emotional interpretation and start to trust muscle memory.

There were two bars in *Les Mis*, not particularly difficult but very fast. I guess I psyched myself when I was practising and played it incorrectly so often that it was already in muscle memory. Undoing it was nigh on impossible. In the end, I just left out those two bars. Lucky it wasn't a solo or an exposed passage!

I wanted to know how all this worked for a string player. Trevor James has spent an entire career playing double bass in the Melbourne Symphony Orchestra. He said that vibration feedback did not apply when playing stringed instruments, despite my assumption that it would. Strings vibrate, don't they?

He assured me that vibration is all in the bowing or playing vibrato, but not for hitting the note. You just have to get your fingers in exactly the right place, often fast. But muscle memory still comes into play:

> The hours spent practising enable me to play the notes without thinking about the mechanics. You might have noticed that the double bass is a big thing. You sometimes have to move your hand by over a foot and put your fingers down in exactly the right spot. That's practising muscle memory. A beginner will get it 1/10 times. A professional should get it 99/100 times and a soloist is expected to get it 999 times out of 1000.
>
> Practising for years gives you the chance of getting it 99/100. More talented people might need less practice than me, but no matter how talented, they still have to practise. It is Edison's 10 per cent inspiration and 90 per cent perspiration.

The musician's description of muscle memory reminded me of my own preparation for card tricks. I have used magic routines in public talks on science and deception and continue to love the art form. When learning a new card routine, I first get the moves right. Once they are committed to memory, I can start playing with the patter and getting the rhythm perfected. My favourite routine involves quite complex manipulations of only three cards. If I need to stop and think about where each of the cards is at any stage of the routine, I will lose the flow and mess it up completely. The rhythm of my hand movements and the story I'm telling are now so closely aligned that the right cards always turn up at the right time. The movements are in muscle memory.

I expect that many readers will recognise the same process within their own areas of expertise. Get the basics in place without error, and then you can add all the glorious creativity and interpretation.

Memorising word for word

Jacqui Dark spent a year touring as Mother Abbess in *The Sound of Music*. She uses what she calls 'pitched vocals' to help with the script. 'When I learn scripts (which I find much harder than memorising music) I create a little "up and down" graph of the pitch of the words in the phrase in my head to help me remember.'

Actor Julie Nihill, who has over 35 years' experience in film, television and theatre, uses a similar technique.

I have noticed, among other things, a certain rhythm association with text. In a script read-through at the commencement of a new project, for example, I've noticed that my body actually flinches when the writing rhythms are 'out'. This is more relevant with a newly written project as opposed to a classic but even then some translations can produce the same effect.

I have also noticed that beautifully written text is extremely easy to learn. I first approach the text by exploring the meaning I am going to give to the lines—this is where all the work is. If the text is well written, I pretty well know the lines by the time I have imbued it with the meaning I have chosen, and physically moved my body with the lines. If the text is 'clunky', or the rhythm's out, then a lot of extra work is required to learn it—and even then, stumbling over words can still happen.

Moving around when studying, creating a rhythmic performance, may well help embed all sorts of knowledge. So why not combine learning with exercise, be that walking, dancing or sweating away on a treadmill? That's what opera singer and actor Brendan Hanson does.

> Last year, I had to memorise an 80-page one-man play. I spent eight months leading up to rehearsals learning it on the cross trainer at the gym, literally walking it into my memory. It had a number of accents that became the music or score for my ear and muscle memory to retain. I trusted that the kinaesthetic part of memorisation, which doesn't normally happen until the play has been blocked, would be covered by the cross trainer. It worked. I draw on the emotional journey as well as a photographic type reference in my mind of the score or script.

All of this talk of rhythms and movement reminded me of the way my Warlpiri colleague Nungarrayi talked about the value of walking while chanting for Australian Aboriginal cultures. The rhythms of the landscape, the chant, the emotional impact and the knowledge form a memorable unit.

When we are happily singing and walking our knowledge, maybe we should be dancing too. One of the most pertinent lessons from indigenous cultures is that knowledge is often danced in ceremonies. I have started adding dance to lots of my songs. Evie Danger is a tap-dancing teacher:

> When teaching dance, it is important to offer just a couple of chunks and weave them together before you add the next chunk. Then you focus on the transition. The interpretation

comes later. Once you know the dance, then you can add self-expression.

Techniques, drills and exercises—you can't get very far into the dance without first focusing on the foundations, otherwise the dance will be jerky and mechanical, not smooth and fluid. You can see it in the faces of the dancers without a firm foundation. They don't look comfortable.

The music is very important. Using the music I love gives the joy, which helps with remembering. I have to choreograph eighteen or so dances a year and without the joy, I can't remember them.

Lisa Minchin took up tap dancing as an adult. She described Evie's class.

The first time on a new dance, my balance is out of whack, because I'm not used to the new combination of movements. I can't think about projection to the audience until I know I won't fall over. It's like making lasagne. You learn the layers, chunk, chunk, chunk. Then you get the pasta to join them. When you cook it, the layers start fusing together. Layer one is knowing where to put your feet. Layer two is controlling your centre of gravity. Further layers add the arms and then the relationship to the music and then facial expression. You only start to live the dance when you integrate the pieces and it becomes natural. The more the dance is fused, the better it is to watch.

In the performance you still have to actually think the really tricky bits: hop, shuffle, step, stamp. But mostly it's in muscle memory and you're free to forget about the steps and just concentrate on the joy of dancing onstage.

There are so many metaphors there for learning—they apply to almost any topic. I am really hooked on the metaphor of fusing my lasagne of knowledge by cooking it well.

Musicians are well aware that the lessons resonate far beyond their own discipline. I asked American jazz performer and author Ted Gioia about the way jazz blends the foundations and improvisation with such wonderful creative flair.

Jazz musicians often learn to improvise by memorizing different melodic and rhythmic phrases and patterns. They use these as building blocks for their solos. When Albert Lord studied the conditions that led to the creation of epic poetry, he found the same procedure. Epic poets in oral cultures know numerous set phrases and these recur over and over again. Homer talked repeatedly of 'rosy-fingered dawn' or 'swift-footed Achilles'—and these phrases were employed because they had the right number of syllables and accents to fill up a place in the structure. Jazz musicians do the same thing with their favorite phrases and licks.

Each player in a jazz group has freedom, but it isn't complete freedom. If I am playing a certain song, there are certain constraints that I can't avoid, but in other aspects I have freedom. The structure and harmonies won't change. The melody has to show up at some point. But these constraints still give me enormous scope for creative expression.

I might compare a jazz musician to a filmmaker who has to turn a familiar story into a movie. There are certain characters and plot elements that must show up in the finished product. But that hardly prevents the director from putting a very personal stamp on the proceedings. That's how Miles Davis played a George Gershwin song, or John Coltrane played a theme from a Broadway musical.

Gioia's description of jazz is the perfect metaphor for the education system I dream about. The knowledge that forms the foundation for any discipline is set. But the stories an individual can bring to it will enable them to take ownership and offer that gorgeous opportunity of 'enormous scope for creative expression'.

Anthropologist Keith Basso described the way Anglo-Americans write Apache tribal history:

> Mute and unperformed, sprawling in its way over time and space alike, it strikes Apache audiences as dense, turgid, and lacking in utility. But far more important is the fact that it does not excite.[1]

Compare this to the indigenous performance of knowledge, bringing it to life.

I have no doubt that we could make almost any school subject more exciting by adding the creativity of a performed knowledge system. We don't need to give up anything we already have that is working well. But we can enhance it with story, song and dance.

Memorising in mathematics

Although people often ask me why we would need to memorise at all when we have Google, they always, almost without exception, comment that students do need to memorise their multiplication tables.

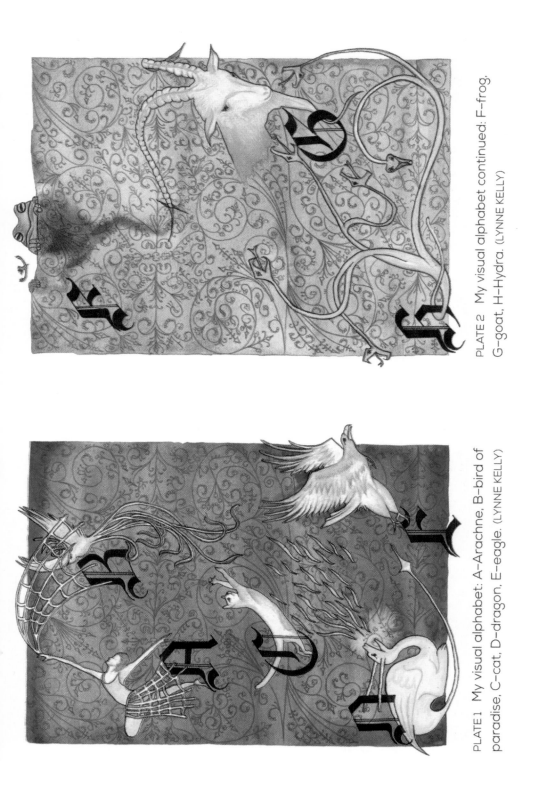

PLATE 1 My visual alphabet: A—Arachne, B—bird of paradise, C—cat, D—dragon, E—eagle. (LYNNE KELLY)

PLATE 2 My visual alphabet continued: F—frog, G—goat, H—Hydra. (LYNNE KELLY)

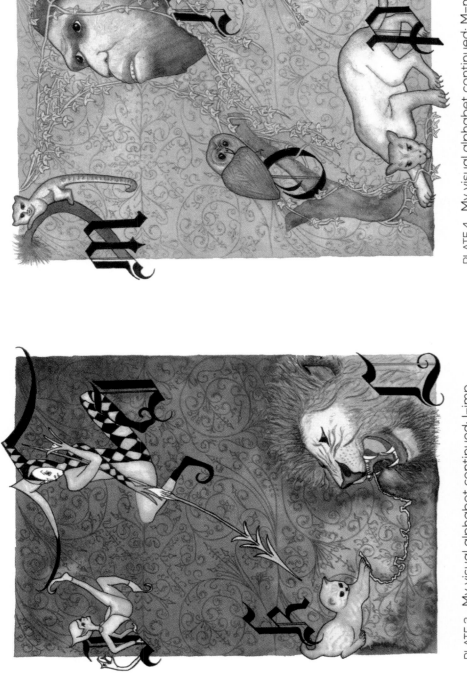

PLATE 3 My visual alphabet continued: I-imp.
J-jester, K-kitten, L-lion. (LYNNE KELLY)

PLATE 4 My visual alphabet continued: M-marmoset.
N-Neanderthal, O-owl, P-panther. (LYNNE KELLY)

PLATE 5 My visual alphabet continued:
Q–Quetzalcoatl, R–rat, S–skull, T–toucan, U–unicorn.
(LYNNE KELLY)

PLATE 6 My visual alphabet continued: V–vulture,
W–wombat, X–Xena, warrior woman, Y–yak, Z–Zeus.
(LYNNE KELLY)

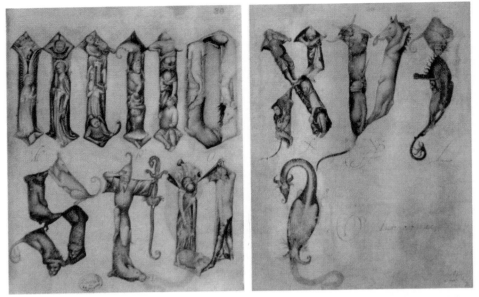

PLATE 7 Two pages from Giovannino de' Grassi's visual alphabet, c. 1390, held by the Biblioteca Civica, Bergamo, Italy.

PLATE 8 Illustration of a bat (folio 51v) from the Aberdeen Bestiary, c. 1200, held by the Aberdeen University Library.

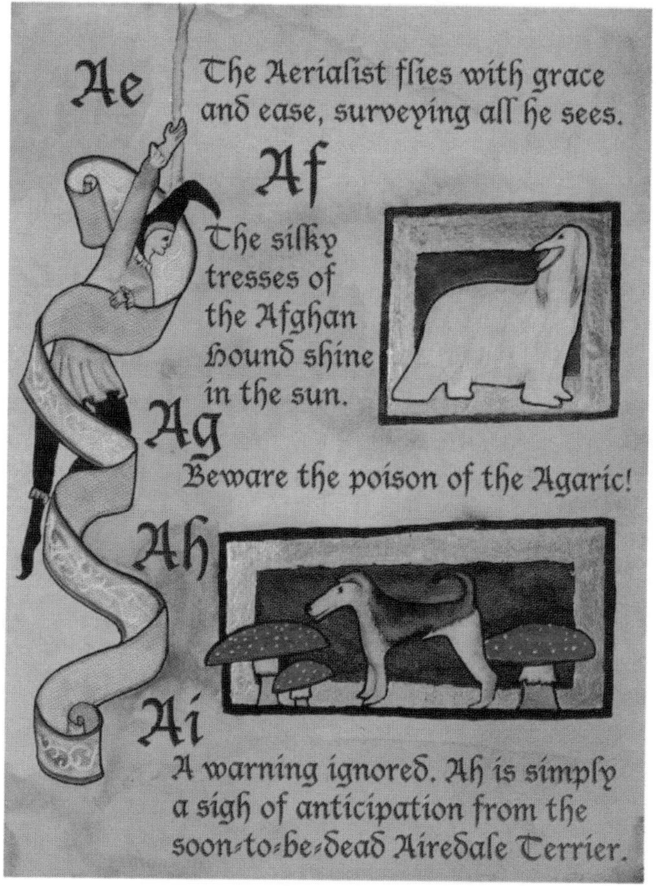

PLATE 10 A sample of my bestiary: Ae–aerialist, Af–Afghan hound, Ag–agaric fungi, Ah–a sigh, Ai–Airedale terrier. (LYNNE KELLY)

Ae The Aerialist flies with grace and ease, surveying all he sees.

Af The silky tresses of the Afghan hound shine in the sun.

Ag Beware the poison of the Agaric!

Ah

Ai A warning ignored. Ah is simply a sigh of anticipation from the soon-to-be-dead Airedale Terrier.

PLATE 9 A sample of my bestiary: Aa–aardvark, Ab–Abyssinian, Ac–acorn, Ad–adder. (LYNNE KELLY)

Aa The Aardvark is proud of his ever so fleshy nose.

Ab The Abyssinian cat is equally proud of her ever so short hair.

Ac The Acorn displays both rough and smooth.

Ad Is this Adder the very snake who gave Adam his apple?

PLATE 11 *Kachina* of the Pueblo cultures from the US southwest, late nineteenth century, as stored at the Brooklyn Museum, New York.
(LYNNE KELLY)

PLATE 12 *Kachina* by Pueblo artist Jacida L, from my collection.
(LYNNE KELLY)

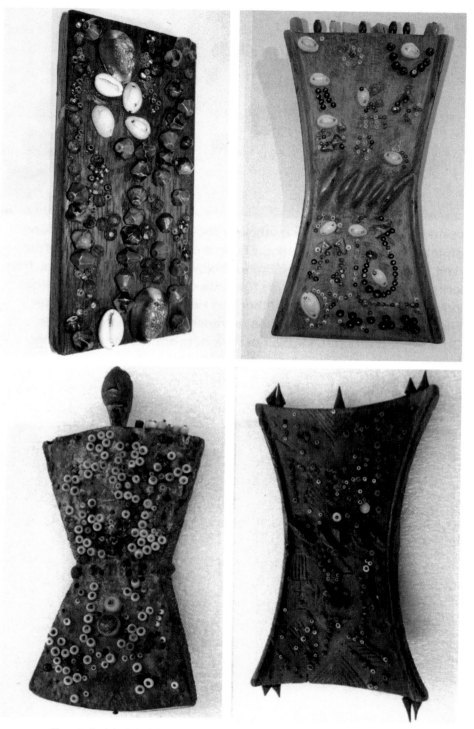

PLATE 13   Top left: My bird *lukasa*. Top right: My *lukasa* for the story of writing. Bottom left and right: two real *lukasas* (or *nkasa*) from the Brooklyn Museum, New York. (LYNNE KELLY)

PLATE 14 Objects for Greek mythology shown centre stage at the end of Act 1 of my personal retelling. Chaos is centre top, with, from left to right, Tartarus, Gaia, Nyx and Erebus below. The next level includes, from left to right, Typhon, Uranus, my favourite stone, Pontus (the markings look like the seashore), then Aether and Hemera with their daughter, Thasalla, below. The lowest stone represents the primordial deities, Ourea. The characters for the next few acts are waiting in the wings. (LYNNE KELLY)

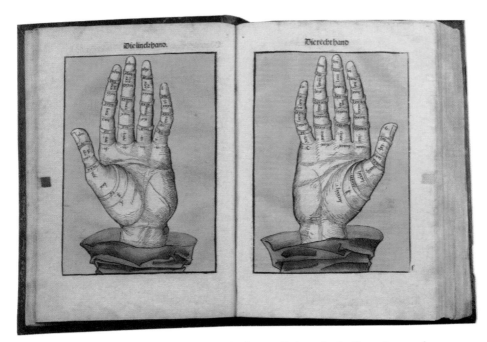

PLATE 15 The *Schatzbehalter* hands, from *Schatzbehalter des wahren Reichtümer des Heils* (1491) by Stephan Fridolin.

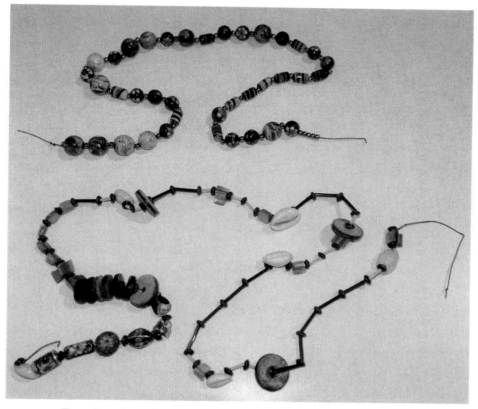

PLATE 16 Top: A string of beads to recall a list of Shakespeare's plays in chronological order. Bottom: The beads representing the main characters and scenes of *A Midsummer Night's Dream*. (LYNNE KELLY)

PLATE 17 *Khipu* in the Museo Machu Picchu, Casa Concha, Cusco.

PLATE 18   Mandala of Vishnu, Nepal, Bhaktapur, dated 1681, currently in the Los Angeles County Museum of Art.

PLATE 19 A Japanese scroll, *Annual Festivities* (*Nenchū gyōji*), created by Ōishi Matora (Shinto) (1794–1833), in the National Gallery of Victoria. (LYNNE KELLY)

PLATE 20 The Rosetta Stone in the British Museum. (HANS HILLEWAERT)

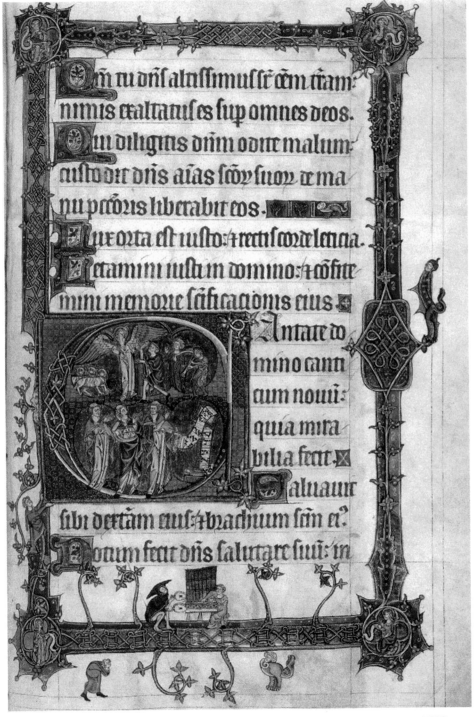

PLATE 21 A page from the Gorleston Psalter, dating from between 1310 and 1324, East Anglia, England. This is just one of the images from the incredible British Library collection available to view online.

PLATE 22 Canon tables from the Book of Kells written in the ninth century CE and held in Trinity College Library, Dublin.

PLATE 23 'Scenes from the Passion', Saint Augustine Gospels. (MS 286, CORPUS CHRISTI COLLEGE, CAMBRIDGE)

PLATE 24 *The Triumph of St Thomas Aquinas*, by Andrea da Firenze, fresco, c. 1366–1367, Cappellone degli Spagnoli, Santa Maria Novella, Florence.

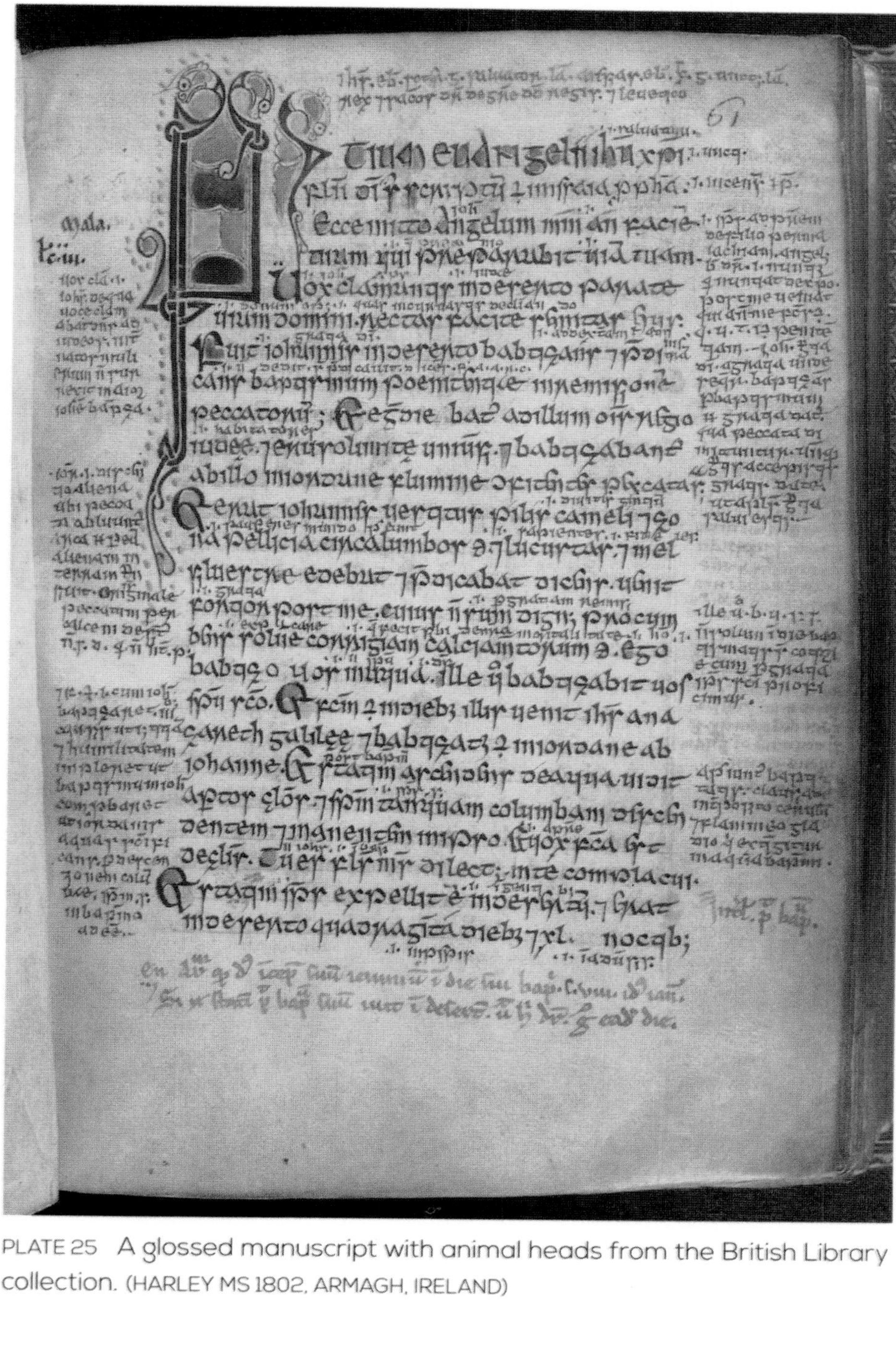

PLATE 25 A glossed manuscript with animal heads from the British Library collection. (HARLEY MS 1802, ARMAGH, IRELAND)

PLATE 26 Geese hanging a fox, one of the many drolleries in the *Decretals of Gregory IX with glossa ordinaria* (the Smithfield Decretals), c. 1340, held by the British Library.

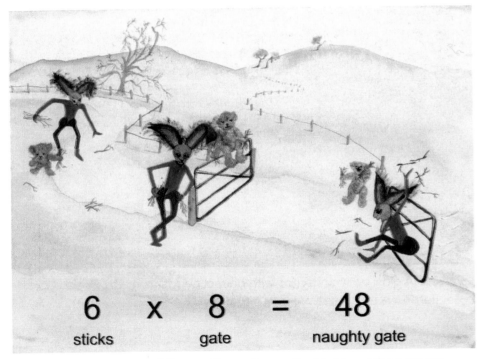

6 × 8 = 48

sticks   gate   naughty gate

PLATE 27 An example of Rapscali's mathematical tables. (LYNNE KELLY)

Back in the olden days, when I was a child, tables were taught by repeating a singsong 'Once one is one, two ones are two, three ones are three . . .' I still know the tune but the words always evade me. I know that it worked for some people, but not for me and not for many others I have asked. When working mathematically, we can't sing through all the tables to get to the one we need. We need to recall any given table instantly.

I developed a plan for memorising the multiplication table from 1×1 to 12×12. I won't explain it in detail here—I'll just outline the method.

The commutative law eliminates nearly half: if you know $6 \times 8 = 48$, then you don't need to remember that $8 \times 6 = 48$. The ones, tens and elevens have easy patterns, each of which can have one story that serves the entire set. Fives and twos can be easily counted while nines can be recalled with a really simple finger pattern. We can add these to our instant memory through stories later, if the need arrives for faster recall; it is more important to get the harder tables memorised.

There are only 31 multiplications that need to be memorised by rote, as shown in the white squares in Figure 8.2. Most occur when the threes, fours, sixes, sevens, eights and twelves interact with each other. Drawing a story requires the student to engage with each multiplication individually. They will end up with a booklet of images for any tables they don't already know in pre-testing. It becomes their reference, but quite soon all the images in it will be familiar and able to be called to mind as needed.

For the numbers, I use the associations 1–sun, 2–shoe, 3–tree, 4–door, 5–hive, 6–sticks, 7–heaven, 8–gate, 9–sign,

	1	2	3	4	5	6	7	8	9	10	11	12
1	1	2	3	4	5	6	7	8	9	10	11	12
2	2	4	6	8	10	12	14	16	18	20	22	24
3	3	6	9	12	15	18	21	24	27	30	33	36
4	4	8	12	16	20	24	28	32	36	40	44	48
5	5	10	15	20	25	30	35	40	45	50	55	60
6	6	12	18	24	30	36	42	48	54	60	66	72
7	7	14	21	28	35	42	49	56	63	70	77	84
8	8	16	24	32	40	48	56	64	72	80	88	96
9	9	18	27	36	45	54	63	72	81	90	99	108
10	10	20	30	40	50	60	70	80	90	100	110	120
11	11	22	33	44	55	66	77	88	99	110	121	132
12	12	24	36	48	60	72	84	96	108	120	132	144

FIGURE 8.2 My multiplication plan. Only 31 white multiplications must be memorised. Those below the diagonal are reflected above it. Multiples of 1, 2, 9, 10 and 11 can be known quickly by pattern. (The 9s can be memorised later by the same method.) (LYNNE KELLY)

10–hen. (I promised you way back in Chapter 1 that I would eventually use that simple mnemonic.) I use 11–legs for the troublesome, rhyme-less 11. And 12–elf is again the best I can manage, although the plural works nicely, twelves-elves.

I also need prefixes for the multiples of ten:

20–plenty of . . . (21 = plenty of sun)
30–dirty . . . (36 = dirty sticks)
40–naughty . . . (42 = naughty shoe)
50–nifty . . . (56 = nifty sticks)
60–fix the . . . (63 = fix the tree)
70–heavenly . . . (72 = heavenly shoe)
80–ate the . . . (81= ate the sun)
90–mighty . . . (96 = mighty sticks)
100–the undead (blame my students for that one)

If the student doesn't know the patterns described above for the ones, twos, fives, nines, tens and elevens, then we create images for the pattern rather than the individual tables. For example, when the rapscallion multiplies by 10, the 10–hen lays an egg (a zero) next to the number. That drawing goes into their reference book to be reviewed if they forget the rule. The more the students have fun with their illustrations, the more the rule stays with them. If they rush them, the experience will be less effective.

For each multiplication they can't recall on a pre-test, students need to make up a story and illustrate it.

In Plate 27, I have represented the table 6 × 8 = 48 accompanied by a drawing revolving around 'sticks' and 'gate' giving 'naughty gate'. The drawing shows my rapscallions picking up sticks, then resting on a gate. One of them is then knocked flying by a naughty gate. The table is written below to reinforce the familiar format. Many students elaborate further by writing stories to accompany their drawings. The student must have time to engage with the story. That time investment will pay off many times over during their life.

Using their own rapscallion makes the stories much more memorable because it brings in a personal and therefore emotional connection. Students become so engaged with their rapscallions that they can't resist the challenge of creating stories. Some students enjoy it so much they create a pantheon of characters or incorporate their friends' rapscallions alongside their own in the narrative.

There are books of table stories given as images that can be given to students as a finished product. These will not be as effective as the students creating their own art with their own rapscallions. That hasn't stopped me painting a complete set of images featuring my rapscallions. How could I resist? It is fun and I had to test the system.[2]

Memorising equations

There are lots of suggestions about how to memorise equations. Unsurprisingly, they generally recommend making a story by giving the letters a character. The easiest way is to have the character name start with the letter in the equation. The characters then interact using the relationships to make a story.

Not all equations are mathematical. At one school, students wanted to remember chemical relationships. For example, 'acid plus base gives salt plus water'. Acid became a teenager who met Base, another teenager, and they travelled to the beach, which led to seaside offspring. These were teenagers; I didn't ask the details of the story but they were all laughing and never forgot the equation.

The more students engage with abstract concepts, act them out and become personally involved, the more these concepts will come to life.

So much to memorise: medicine and law

It's not just school students who need to memorise the foundation of their subjects. The two university courses considered to have the highest dependence on memory are medicine and law. In both cases there is plenty of evidence of the value of using memory palaces associated with the most common techniques: vivid imagination and stories.

There must be few people in the world better qualified to offer advice on using the memory methods for academic study than world memory champion Alex Mullen and his partner, Cathy Chen. They run a great website, Mullen Memory, on all things memory related.[3] Mullen can memorise the order of a shuffled deck of cards in under twenty seconds, among numerous other phenomenal feats. At the time of writing, he is the top-ranked memory athlete in the world.

Chen and Mullen both have degrees in engineering, and Mullen is also qualified in mathematics. They are both studying medicine at the University of Mississippi, so it should be no surprise that the topics they use to demonstrate the memory techniques are anatomy and pharmacology.

They emphasise that stories, imagination and memory palaces only serve as part of a medical student's training. Chen and Mullen don't advocate memorising anything that can be intuitively understood. Mullen doesn't set up his memory palaces in advance, but creates them as part of the process of learning. In this way they are optimised to the actual material he's studying. Here he gives an example:

Let's say I'm learning about antibiotics, and at the start of the chapter I decided to use my elementary school as my palace.

As I learned, I followed a path organically through the school, choosing rooms/areas to neatly house my images for each chunk of info. Let's say I've just used the school lobby as my dedicated area for the drug class tetracyclines. Now, the next chunk of info being about the drug chloramphenicol, I might choose the room just right of the lobby to house all my high-yield chloramphenicol images, selecting loci within that room to fit the facts. This approach allows me to choose rooms/areas/loci that most appropriately match the material. In this way, the creation of palaces fits quite seamlessly with my own analysis of what's worth memorizing and how I should structure it, so I find in fact that it aids in my critical thinking about the material. It also cuts down time spent on 'palace building'.

When I have designed a memory palace or portable device at the same time as engaging with the information, it has worked as seamlessly as Mullen describes.

Mullen uses spaced repetition systems; he particularly recommends the extremely popular SRS software Anki. I suspect that if I were studying for university exams, I would likely use the software as well.

Like medicine, studying law requires a strong factual foundation that can be adapted to new situations. The legal profession constantly requires new knowledge and understanding, new arguments and applications that are not possible without knowing what has gone before. Details can be looked up online and in journals, but a basic knowledge of the law is essential in order to know what to look up, where and why.

The legal profession has been memorising a whole swag of factual points, laws and precedents for millennia. I described the mandala of the Eddie Mabo case in Chapter 6 and the gloriously decorated medieval Smithfield Decretals in Chapter 7. Long before both, Cicero was writing about using memory palaces for conducting legal arguments in court.

Cicero was not only known for his phenomenally trained memory—his fame as an orator was often witnessed in court where he practised as a lawyer in the first century BCE. Cicero wrote that he employed a large number of places in his memory palaces. Each location was well lit, clearly set out in order, with moderate intervals between, and he used images that were active, clearly defined, unusual and fast to recall. Sound familiar? It should, because Cicero was drawing his method from that famous textbook the *Rhetorica ad Herennium*. There are very few examples in the *Ad Herennium* because the anonymous author advised that images made up by others are nowhere near as effective as your own. But he did provide a legal case.

The defendant is charged with killing a man by poison to gain an inheritance. There were many witnesses. Our anonymous textbook author suggests that you, as the legal counsel for the defence, should imagine the victim lying in bed with the defendant beside him. The defendant holds a cup in his right hand and stone tablets in his left. That much makes sense: the cup refers to the poison and the tablets to the will. The author also suggests that on the defendant's forefinger we imagine a ram's testicles, which is not quite so intuitive. The Latin word '*testis*' means testicles, and also refers to a witness. The teacher highly recommended verbal playfulness as a method for imagining abstract concepts. It worked then and it still works now.

As the legal representative, you should store further details in locations around the memory palace using images. You need to memorise the facts of the case, legal precedents and the arguments you are going to use in court. The palace ensures that you will argue your case systematically, location by location, as planned.

Education doesn't stop after university. It goes on throughout life, preferably a long life with an active brain. Our memories are our identity. If we keep learning, keep memorising, keep laying down and reinforcing our neural networks, can we delay or even halt the dementia people start dreading as they approach old age? I sincerely hope so. Time to see what the research says.

CHAPTER 9

Does memory have to decline when you age?

Memory loss in old age is inevitable, isn't it? The answer is no—it's certainly very common, but not inevitable.

If you've read this far in the book, you'll know I am going to wax lyrical about memory training for older people. I strongly believe that there are ways of targeting that training to retain useful information and, in particular, plan for retaining a sense of self.

At almost every talk I give, someone with greying hair will ask if memory training will prevent dementia. And let's be frank, I'm 67 at the time of writing so it's a pretty big issue for me too. I would love to say, 'Yes! The evidence is all there. Get training!'

Unfortunately, the evidence to date is inconclusive. But there are *strong* indications that any kind of increased and intense cognitive activity gives you a much better chance of at least delaying the onset of dementia, if not preventing it altogether. That's good enough for me.

Exactly what sort of training would be best? And are there other strategies that might help in retaining identity and cognitive function even once dementia has been diagnosed?

I have been intrigued by a number of documentaries and news snippets about dementia over the last few years. They show almost unresponsive people suddenly stand up and dance when familiar music was played. Some, who apparently no longer spoke, even started singing the words.

In other cases, dementia patients were taken to places they knew well before the disease had struck, particularly places from much earlier in their lives. The patients not only recognised familiar sights, but responded really strongly to the experience. Others responded well to a toy bear or doll. Given that music, place, dance and non-human characters have proven to be so valuable in memory work, isn't this a major clue to what we can do throughout life that may well delay the onset of dementia, or at least reduce its impact?

The ideas presented in this chapter are a little speculative, but it is rational speculation backed by a fair bit of hard-core evidence. Everything I suggest can only be beneficial; there are no negative side effects to memorisation. And it's free.

Is memory loss normal?

Dementia affects almost 50 million people worldwide, a number predicted to increase rapidly as baby boomers age. Dementia is the leading cause of disability in older people, but many don't realise it is also one of the leading causes of death.

This is a devastating disease for any family, both emotionally and financially.

It is important to recognise the difference between dementia and the occasional lapse of memory. Researchers differentiate on the basis that normal memory deterioration does not have any major impact on your life, while dementia becomes disabling. Normal memory deterioration may not entail more than the occasional senior moment, while dementia will get worse, often quite quickly.

I have noticed that older people attribute lapses of memory to their age, while younger people simply laugh off the odd occasion they go into a room and can't remember why. Older people get concerned when they forget where they left the keys, while younger ones assume they were simply thinking of something else.

Nevertheless, the research is fairly clear: older people will gradually find remembering more challenging. But is this just a case of 'use it or lose it', like any muscle? Or is it an inevitable side effect of ageing? I'm sorry to say I don't have a definitive answer, but I do suspect that the expectation that memory loss is normal has stopped many older people from actively exercising their memories.

I wonder if our increasing propensity to rely on Google rather than remember anything is causing these increasing rates of dementia. Are we taking the easy way out mentally, and consequently taking no advantage of lifelong brain plasticity? By using our brains so much less we're not forcing new neurons to lay down fresh pathways and enforce those already there.

We must aim to retain a lifetime's learning for as long as we can. But we also need to recognise our family and friends—to know who they are and remember the past we have shared with them. Our identity is enmeshed with the relationships that have defined our life.

We also need short-term memory—the working memory that is only stored for seconds, at most minutes. Those snapshots of life, sensations, emotions and information, in their wondrous assortment of formats, will disappear if they're not cemented into some form of long-term memory. We need to create and reinforce neural pathways, keeping them open to all the information stored in these incredible brains of ours.

What is dementia?

Dementia is a neurodegenerative disease. That means that it is caused by the progressive deterioration of nerve cells. Sometimes the cells die completely. Dementia is not one specific disease but a range of disorders. Symptoms vary between individuals, but essentially, dementia affects thinking, behaviour and the sufferer's ability to engage in work or social life in the way they had previously. It eventually interferes with their ability to perform everyday tasks and live an independent life. Dementia is not an inevitable part of ageing. A significant proportion of the older community do not develop dementia; meanwhile, there are occasional cases of younger people suffering from this disease.

It is not surprising that studies on dementia are consistent with the neuroscience of memory described in Chapter 3. With Alzheimer's disease, patients exhibit damage to neurons, particularly in the hippocampus, which stops them forming new memories. Short-term memory becomes impaired. The damage also messes up neurons in the networks that encode memories from before the disease took hold, known as retrograde amnesia. It is not just the connections that are lost—with time, the physical regions of the brain actually shrink and waste away. With advanced Alzheimer's, a great deal of brain tissue has been lost. Memories are lost, and then language, and then eventually motor skills. Not a pleasant prospect at all.

Family members commonly say the most distressing aspect of dementia is the loss of identity. Patients often lose the ability to recognise even those closest to them. If there is anything at all we can do—for those with dementia and to prevent it in others—and if there are no negative side effects, then nothing should be stopping us.

Clearly, I am in no position to discuss all the medical treatments currently in use for dementia patients. My goal is to look at the role memory methods could play, given that they relate so closely to neuroscience, have been in use for millennia and loss of memory is a huge part of these diseases.

Memory palaces and dementia

There has been much research into the therapeutic use of memory palaces in dementia. In one study, conducted in 2010, a Norwegian team from the University of Oslo put volunteers with an average age of 61 through an intensive program for eight weeks. They asked volunteers to use the method of loci—that is, memory palaces—to memorise a list of words. Among other positive effects, after the eight weeks structural MRI images showed a surprisingly impressive thickening of areas of the brain in the cerebral cortex: in other words, a demonstration of brain plasticity in ageing volunteers, always encouraging for someone of my age.

The cortical thickening was found to be directly proportional to the volunteer's improved memory performance. The control group showed a trend towards thinning in their cortices over the same time, which may be a natural part of ageing without any memory work. The researchers concluded that the method of loci leads to long-term memory retention even after initial training.

I found this study particularly intriguing. As I'll describe in the next chapter, one event in memory championships is exactly this: memorising lists of words in order. I train for this event regularly, so I really want to believe this research! The authors warned that these were early results with a small sample, but I have since found quite a few reports with similar findings, and none disagreeing.

There are studies looking into the use of memory palaces for people with memory problems caused either by brain damage or age. The research showed that visual memories could be created even when other forms of memory were a real struggle. Creating visual memories is the basis of all the memory methods described in this book. I feel really confident recommending that you get out there and create memory palaces for the sake of your long-term memory.

There has been quite a lot of experimentation with creating virtual memory palaces—nice computer-generated buildings and landscapes to save you having to walk anywhere. They have been shown to be very successful in a huge range of contexts. You can make them more dramatic and interesting than the real world. Some people just like them better. I like my real palaces and don't respond emotionally to virtual palaces. It's just a matter of personal taste. However, virtual palaces are obviously advantageous for anyone who finds moving around the streets difficult. This is often the case for the elderly. Dementia patients have a specific set of needs that virtual memory palaces can address in a way that my beloved physical palaces fail to do.

Researchers from KU Leuven, a university in Belgium, decided to address the issue of recognising faces. Dementia patients get very distressed when they are unable to recall the faces and names of loved ones; it is also very upsetting for family and friends. The researchers are convinced that memory training is a promising intervention for geriatric neurology problems.[1]

The researchers designed their experiment around the neuroscience that indicates that both the method of loci and memorising names and faces use brain structures that are fairly well preserved as dementia progresses. Caregivers constructed virtual scale models of the homes of early onset dementia patients, which functioned as memory palaces. Photos of people who were significant in the patient's life were placed around images of a pathway through the home in which they lived before they went into care. Patients would then 'walk' their familiar environment with a caregiver or family member, to recall faces and associate names. In this way, the patient took their virtual home with them into their aged care setting.

The patients were prompted to tell positive stories about the significant people in their lives. This helped caregivers learn about the patient, adding a more personal element to their care than would normally be the case from clinical notes. The researchers have reported that the initial results are promising. A reduction in depression was also indicated, a pretty great added bonus.

This is but one of many encouraging research projects using memory palaces to alleviate memory loss.

The power of music and memories

The documentary *Alive Inside* caused a massive reaction globally: filmmaker Michael Rossato-Bennett, along with Dan Cohen, founder of the non-profit organisation Music & Memory, recorded dementia sufferers all over the United States responding strongly to personalised suites of music. Rossato-Bennett described the music as not only combating memory loss but restoring a deep sense of self to sufferers, and carers around the world have replicated these findings. It seems that the part of the brain that responds to music may be in an area not as harshly affected as dementia takes hold.

Other research has found positive effects of music therapy on maintaining and even improving cognitive skills. At the same time, the music makes those with dementia more social and active. Further research suggests music reduces behavioural problems, and gives family and friends the tools for a shared pleasure.

The evidence that music is a powerful memory aid is irrefutable. And so I wonder whether we should create a personalised music suite as part of our preparation for old age? And not just favourite music, but why not also songs that reinforce critical information? These could be songs telling of your family, or songs recalling significant events from your life. They could use familiar tunes with the words changed, or be songs specially written for the purpose.

How many older people will still respond to the nursery rhymes they learned as young children? How many reading this book can still sing all those childhood tunes? Clearly the human brain responds to music in a vital and valuable way.

Prevention is better than a cure (which doesn't yet exist anyway)

There is no known cure for dementia. But there are lots of suggestions for preventing, or at least delaying, its onset. Nearly everything I have read quotes research suggesting that a healthy diet, plenty of exercise, quitting smoking, reducing alcohol, getting plenty of sleep, reading, increasing social interaction and reducing stress are all important for dementia prevention, as they are for almost any other disease you care to mention.

Dementia organisations and researchers talk about a range of 'cognitive preventions': the impact of the pathologies associated with dementia can be reduced by building up your 'cognitive reserve'. You start with more cognitive reserve if you have a higher education level, but it can be built up by anything that gives your brain a serious workout.

There are a wide range of programs, all over the world, targeted at those with early signs of dementia, those worrying about it and those who just want to maximise their mental capacity. Dementia organisations around the world suggest doing puzzles like Sudoku and crosswords. Some recommend playing games like Brain Metrix, CogniFit, Dakim Brain-Fitness, Eidetic and, the most common, Lumosity. There are even games designed for those already with Alzheimer's, such as Clevermind.

But are these interventions addressing identity? Can what I am suggesting do all they do, and more?

Lots of these programs involve using memory palaces in some way, but I still found that most of them were about training without a personal context. They play around with information that is of no use beyond the training sessions, such as memorising lists of words or brain-training games. I've no objection to any of these but wonder if we can't do something a little more practical while still exercising our brains to the same level.

Looming large among the recommendations for brain training is learning languages, as described in Chapter 3. Placing history in a memory palace, especially if adding dates, really pushes the brain, as will almost any of the experiments described in previous chapters. But what about the issue of identity?

Dementia and identity

When I talk with dementia patients and their families—and with symptom-free middle-aged and older people who are worried about the disease—I hear a litany of fears, but the most common is the loss of identity. Family members often describe the dementia patient as 'no longer the person they were before'. Your identity is your name, your past and your ability to remember those past events. We identify so closely with the roles we have during our life—in our careers and in our family—and dementia steals that identity.

Much of the research I have read has emphasised the importance of carers trying to help sufferers retain as much self-identity as possible. Carers, at home or in residential facilities, are encouraged to closely engage with the life story of the patient. Experts suggest recording life stories, and including not only careers and occupations but hobbies and favourite activities. They emphasise the importance of personal accomplishments, of memories from school, where they lived, favourite possessions, memorable holidays and experiences within the community or religious affiliation. And of course, the importance of social relationships with families and friends.

What if we prepared for this in advance with a memory aid we could take with us? What would the impact be if we prepared for later years by creating songs containing valuable information and embedding knowledge in familiar places? I suspect that the impact of memory loss and ageing would be reduced. Portable memory spaces can be kept by the elderly, even if they move home or into residential care.

Pictures of loved ones are often used as a memory prompt, but I have in mind something a little more esoteric: a movable memory palace based on Native American winter counts.

A winter count for your life

The Lakota tribes, among others, create 'winter counts'. These are animal hides, or other large drawing surfaces, on which a new symbol is added each year. The symbol represents a key event from the previous year, added on the first day of winter; that is, the day of the first snowfall. The keeper of the winter count is also the keeper of the tribe's history. He or she can recite the stories from the past using the winter count as the guide. The tribe often keeps a number of copies of their winter counts, as is the case with the most famous, that of Lone Dog, shown in Figure 9.1. It begins in the centre at the year 1800 and spirals out to 1871.

FIGURE 9.1 Lone Dog's winter count records 70 years of memorable events for the Lakota Sioux. (COURTESY OF THE MUSEUM OF NATIVE AMERICAN HISTORY, BENTONVILLE, ARKANSAS)

One of my colleagues, Alice Steel, has run workshops for people of all ages to create their personal winter counts. The participants delighted in designing an artwork representing events of their life to date. Their winter count was then ready to be extended by one symbol per year. If they follow the Lakota tradition, that would be on the first day of winter, although waiting for the first snowfall has some obvious disadvantages for Australians. Some people chose the date of the official start of winter while others chose to commence their

FIGURE 9.2 Alice Steel with her son, Haku, telling stories from Alice's winter count. (ALICE STEEL)

winter counts from their birthdates or New Year's Day. They still used the term 'winter counts' to remind them that the inspiration came from the Lakota. People recalled and encoded really revealing and rewarding life stories.

Alice has explained the process for constructing her winter count, shown in Figure 9.2.

There are 36 pictograms on my winter count, spiralling out from a central symbol that marks the day I was born. Each symbol denotes a year of my life. They have been carefully designed to represent my most significant memories. I have laboriously embroidered them onto a small square of canvas, so even with my eyes closed I can feel all the stages of my existence. I long to add the next symbol but a new winter has not yet arrived.

My winter count also contains all my shared memories with the important people in my life. It connects my history with theirs. Such a simple little object and yet it is now one of my most prized creations. I refer to it all the time when figuring out dates from my past, even when filling out convoluted government forms. Having my life displayed all on one page also provides an interesting perspective. I've noticed patterns that repeat. Events, many years apart, that are curiously connected.

I watched the process from the sidelines. Alice lives next door to her mother, who often takes care of Alice's young son, Haku, a preschooler at the time. The three generations became immersed in developing their winter count together. Alice explained:

Photographs provide a good chronological reference for some memories but as many of them are just people smiling at the camera, they rarely depict the actual events from my most vivid memories. It was much more useful talking to my family and friends. I asked my mother endless questions, going through her old diaries and even arguing over things we remembered differently. She was essential for information about my very early childhood, which I could not recall in so much detail. This process actually woke up a lot of dormant memories for her too, and she is now very keen to make her own winter count before the memories are lost.

My son, Haku, was three years old when we started. He is fascinated with my winter count. He loves that he can point to any of the symbols and make Mummy tell another story.

He is starting to recognise which symbols belong to which stories. He loves the ones that are from when I was a child and the stories that include him. For now, he still wants me to tell the stories but I imagine one day I will need him to tell me. Soon, I think, he will be ready for his own winter count.

You've probably seen Grandma or Grandpa start to recount a story from their life only for grandchildren to stop them with an 'I've heard that one'. With a winter count, the story-teller role can be reversed: children can point to the icons and relate the stories back to their elders. The embroidered or illustrated cloth becomes an heirloom for the future. And it is much more fun for the children.

Alice also reflected on the value of a winter count when recalling her grandmother's last years.

Towards the end of my grandmother's life her memory began to fade noticeably. Any time she had to go into hospital her memory would deteriorate rapidly, yet it would improve again once she returned home. It was clear how much her familiar environment and possessions tied down her memories. When she was removed from these, her already fragile mind was untethered, leaving her very confused and disoriented. Having something easily transportable that holds all your most important memories could be a real lifeline for people in these situations.

Researchers point to the problem with dementia patients in residential care becoming completely absorbed within their disease and the hospital. If a patient had a personal winter count, the therapist could use it to prompt chatter about their life and their symbols. But a word of warning: these are not easy to construct, so waiting until someone has dementia will be way too late.

In Alice's workshops, winter counts were drawn with marker pens on rolls of canvas in preparation for conversion to any material.

I asked them all what their earliest memories were. A common statement was that they weren't sure if it was truly their own memory or something they were told about their childhood. They could then see how important it was for them to provide this for their own children. Some of the participants were also interested in possibly creating a collective winter count for the whole family.

They also had some ideas to experiment with the design. 'Does it have to be a spiral?' Others suggested tapestries or quilting. I didn't see why not, as long as they could distinguish the order of years and identify each significant memory.

I found creating a personal winter count for 67 years very difficult. So many years have passed that have merged into sameness, although they certainly weren't so at the time. Often, I have to extrapolate from the age of my daughter in any given year to work out what I would have been doing. Last year's symbol represents the first time I competed in a memory competition. I am over 60. I am a 'senior'.

The number of seniors competing in memory champion-ships is small, but growing. Japanese Hiroshi Abe is a year older than me. He competes both nationally and internationally, while Marie Kaye first entered a US competition when she was 70.

Dominic O'Brien has eight world championships to his name. He won the 2018 United Kingdom and Pan European Memory Championships, beating all the youngsters. He is also the World Senior Memory Champion, at 61. He is on a mission to convince the world that we do not need to lose our memories with age: 'It is not age that makes the difference, but how you challenge and stretch your brain every day.' He believes that the best way to do this is to take part in competitive memory sports.

So what exactly are they?

CHAPTER 10

Memory athletes battle it out

I said I wouldn't do it. I swore I wouldn't train for memory competitions, sitting at a desk for hours wearing earmuffs and a frozen stare, going over and over shuffled decks of cards and long lists of random numbers.

I watched the 2016 Australian Championships in Melbourne. This is no spectator sport: the twenty or so competitors sat at desks in silence, headphones on and eyes down. During their ten events, they barely stirred. Any of the very few spectators who budged or even breathed noisily received a scathing glare from the arbiters.

I walked away in awe of the competitors' ability to concentrate and memorise vast sets of boring, useless information—and also sure I was incapable of handling the pressure of two days of intense concentration and competition.

But how could I write a chapter of this book on memory competitions without having experienced them myself? I dithered and delayed but have noticed so often that understanding is so greatly enhanced by participation, that I relented. At the age of 65, I started training. When I met Dominic O'Brien—the most successful memory athlete in the history of the sport—I seized the opportunity to seek his advice. I started memorising cards.

I was willing to suffer for you, dear reader, but I found myself enjoying getting better and better at dreaming up bizarre images on the fly. I was a little shocked by how raunchy and violent some of them became, and even more surprised by how rapidly they started to form with practice. It is impossible to describe my elation the day I first memorised an entire deck of cards. It took me 37 minutes—a tad longer than the current world record time of 12.74 seconds—but I had done something of which I had been convinced I was incapable.

So I entered the 2017 Australian Memory Championships. And I surprised myself: I could handle the pressure. I took out the Senior Memory Champion title and retained the title at the 2018 competitions.

The disciplines

The same rules apply across the world for memory competitions. Speed Cards is the glamour event and always held last, for the drama. There are two trials. World records reflect the international event lengths. At both national and international memory championships there are ten disciplines, the only difference being the time allocated to each:

1. Shuffled decks of cards (national: 10 minutes; international: 1 hour)

2. Random Numbers (national: 15 minutes; international: 1 hour)

3. Speed Numbers (national and international: 5 minutes, best of 2 trials)

4. Binary Digits (national: 5 minutes; international: 30 minutes)

5. Spoken Numbers (one read out per second; as long as you can manage, 2 trials)

6. Fictional Dates (fictional; national and international: 5 minutes)

7. Names and Faces (national: 5 minutes; international: 15 minutes)

8. Random Words (national: 5 minutes; international: 30 minutes)

9. Random Images (national: 5 minutes; international: 15 minutes)

10. Speed Cards (time for a deck up to 5 minutes maximum)

The US championships have a slightly different format for their final. It is performed on stage in front of an audience, and the events include memorising an unpublished poem and recalling details about people who announce their data on the stage. It sounds incredibly stressful.

Most events require the competitors to use memory palaces to store and recall the information. I keep the memory palaces I use for competition empty of any other data, with one exception: for the Random Words and Random Images events, I use my Countries Journey. For memorising the countries, I use many of the features of the location: a fence might help remember the capital while a window is associated with some aspect of history. As Dominic advised, for competition I focused on a single aspect of the location, the solar panels on the roof, a chair on the balcony or a fork in a tree.

Some of the palaces I created in training for memory competitions worked extremely well straight away—my Library memory palace, for example. It starts at the entrance to the gym on the corner of my street and visits the town hall, the community garden and an elderly citizens building. Then it goes into the library, around the foyer, past the theatre, up the ramp and into the newspaper reading room. There's a chat at the desk, a pop into the children's room and staff room, a moment to check out the seating areas and the book locations. It finishes with a look out a window.

Other palaces were more of a struggle to establish because there wasn't enough variety. Dominic had warned me against using streets of houses or rows of shops. These work fine when you can ponder the palace at leisure, but when you need to dash through the locations at speed, you need variety. I learned that lesson slowly, eventually giving up on some palaces I'd tried to establish. But the Library memory palace remains one of my favourites.

Memorising a shuffled deck of cards

The way to memorise a deck of cards is to turn each card into a character, to give it life. The Seven of Hearts is just one card among many, but make it Attila the Hun and it becomes memorable.

Many memory athletes assign characters to cards, with a system for the suits. For example, they may use hearts for family members, spades for famous leaders, diamonds for actors and clubs for sports stars. I would have saved a lot of time had I done the same.

I had already attributed characters to my cards because I had used a deck of cards to represent my 'ancestors' long before I decided to compete. But then I struggled to recall some of them as rapidly as I needed to during competitions. Homer, Herodotus, Cicero, Caesar, Plato and Socrates all just blurred together into men with beards and togas. If I had set up the card characters purely for competition, I would have chosen more wisely. I solved the problem by adapting most of the characters. Socrates alone kept his toga.

Homer Simpson stepped in for his more illustrious namesake. (Abusive emails from fans of *The Simpsons* will be deleted without reading). Aristotle became drunken 'Tottle, bloodied from his falls, while Thomas Aquinas started aqua-planing on top of a water spout. Others just got a little more violent or vulgar. William of Occam became a homicidal slasher with his razor, while Plato took to 'plate'-glass windows with a sledgehammer. Attila the Hun became Attila the Honey, a bee-encrusted dude who trotted around swatting his six-legged mates. I had no trouble converting these caricatures back into their historical equivalents when on the History Journey or talking to my ancestors.

With my melodramatic entourage beside me in the 2017 competition, I managed to memorise one deck in about eight minutes. I started on a second deck, but messed it up. Although I scored nothing for the second, I had the entire first deck correct and scored 52 cards. I was absolutely delighted—I had been sure I would falter under the pressure. I managed 73 cards at the 2018 competition.

The world record is currently held by Mongolian teenager Munkhshur Narmandakh, who memorised 1924 cards in an hour. A little easier to visualise, but just as difficult to comprehend, is that she memorised 37 decks of cards. Memorising a shuffled deck of cards means placing each of the card characters in an empty memory palace, with 52 locations giving the order of the shuffled deck.

For example, using my Library memory palace, I need to place the first card at the entrance to the gym. If the Seven of Hearts turned up first, I needed to imagine Attila the Honey on the steps leading into the gym. If the second card is the Ace of Spades, then Homer Simpson is entering the Town Hall. When I need to recall the shuffled deck of cards, I go for an imaginary walk to the Library, collecting each character—and therefore each card—in order.

Adding an action and object to your person

Fifty-two cards, one at a time, require 52 memory locations, but to reduce the number of locations needed, memory athletes grant each character an action. Socrates, for example, drinks poison. Homer eats doughnuts. Attila swats his bees. Cleopatra (nicknamed Kleo, the King of Clubs) is bitten by a snake, Henry VIII (Three of Clubs) decapitates his wives, while Occam (Two of Diamonds) slashes everything in sight with his razor. Stereotypes work a treat, even if historically inaccurate.

Two cards can now be placed in each location by taking the character of the first card and assigning him or her the action of the second. If the King of Clubs is followed by the Ace of Spades, then Cleopatra eats doughnuts. If the Seven of Hearts is followed by the Three of Clubs, then my bee-keeping Attila decapitates his wife. This reduces the number of locations needed to 26.

The People–Action–Object system (PAO) enables *three* cards to be placed together, reducing the number of locations needed for an entire deck to 18. For each character, an action and a separate object are assigned. Henry VIII decapitates (his action) and also holds a decapitated head (his object). Cleopatra is bitten (her action) and has a snake (her object). Homer's action is to eat his doughnuts, while Socrates drinks his poison.

So the King of Spades followed by the Ace of Spades and then the Two of Diamonds means that Cleopatra (the person) is required to eat (Homer's action) razorblades (Occam's object). Things can get rather gory, and therefore memorable. If the sequence is Ace of Spades, Three of Clubs, King of Spades, then Homer decapitates a snake. I can decapitate snakes and bees in my imagination although the thought horrifies me in reality. I need to imagine decapitating a glass of poison or a shard of glass if the Five or Six of Spades (Socrates or Plato respectively) appear in the third position. That just requires broadening the definition of 'decapitating'.

There is an unimaginably huge number of possible orders for a shuffled deck of cards. It is an 8 with 67 zeros after it; that is, 80,000 billion billion billion billion billion billion billion possible combinations. No one would even consider memorising every possible order in advance of competition.

There are even 132,600 different ways of drawing three cards, if the order matters, as it does for the PAO system. No one is going to attempt to remember all those either. So you have no choice but to make the image on the fly for every combination as you sight it. You do get better with practice. Sometimes a really great image slows me down as I ponder it or start to laugh. That's not a good idea in competition.

A haunting fear of ghosts

Memory athletes are all scared of ghosts. One day I went for a walk along a route that takes me through a memory palace I had used that very morning for training. Only a few hours earlier, I had struggled to find a reason for Alexander the Great (Eight of Spades) to be riding his elephant on the tennis court, shining a torch (Six of Hearts) on a razor blade (Two of Diamonds). Ah! He was there to stop the tennis players cutting their feet. Now, on the afternoon walk, that image was still there. Had I allowed myself to concentrate on it, even momentarily, I could easily have fixed the image permanently—Great Alex would have haunted those tennis courts forever. Every time I used the tennis courts for training or in a competition, the ghost of Alex would have raised his ugly head. I passed through that memory palace with eyes fixed firmly on the ground and swore never to walk a memory palace after using it in training ever again.

More common is getting an image from a previous session when you reach a particular location in recall. It is why memory athletes have multiple palaces, so there are always a few days between using any particular location. When you are heading into a competition, the palaces you will use are kept pristine for a week beforehand, or elephant-riding Alex might insist on shining his torch on a razor blade at the tennis courts when he was supposed to be drinking poison in my neighbour's garden.

The events involving numbers use a similar technique where those ghosts cause just as much hassle.

Memorising numbers

Five memory competition events require memorising numbers. This skill is useful in so many aspects of life that I use it constantly.

For the Random Numbers and Speed Numbers events, rows of 40 random digits are presented. The task is to memorise as many of these as you can in the time allotted. The first error halves your score for that row, a second means you score zero for that row.

The first few rows might look like this:

2143223669431671869996137544436484647715
8373928266266395581296887528346271454855
7576471171638986614496635547228943584243
8794747536489396528635762366664997963117

There are a number of ways to memorise long numbers. The two most commonly used are the Dominic System (as explained in Chapter 4) and the Major System. I tried both and the former was more effective for me.

There are lots of different versions of the Major System, but they are all pretty similar. First you assign sounds to the numbers.

0: 'z', 's' or soft 'c', which all sound like 'z', the first letter of zero.

1: 't' or 'd' because they sound similar and both have a single downstroke.

2: the letter 'n' has two downstrokes.

3: 'm' has three.

4: 'r' is the last letter of four.

5: 'f' or 'v' because they are both in five.

6: a soft 'g'—and other letter combinations which sound like it, such as 'sh', soft 'ch', 'j', 'zh'—because a 'g' looks a bit like a flipped 6.

7: 'k'—and sound-alikes such as hard 'c', hard 'g', hard 'ch', 'q', 'qu'—because a capital 'K' is composed of two sevens on their sides.

8: 'l' because both 'l' and '8' are symmetrical and upright.

9: 'p', because it's a mirror image of 9, and 'b' because it's a rotation of 9.

You can add a, e, i, o, u, w, h and y anywhere you like to make the sound of words and so visualise the image. The number 32, for example, would give you 'm' and 'n', so 'man' is the obvious image. Some take a bit more work, but you think them all through in advance and just use the same image over and over. They become your set of 100 mnemonics for the numbers 00 to 99.

You can go on to encode 100 to 999. For example, the number 201 could become 'r' + 's' + 't' . . . and a roast dinner is brought to mind. After only a few times the letters will become enough to suggest the word you have chosen, and you'll get the image pretty quickly.

I use the Dominic System in exactly the same way I do cards. I have a character for every pair of numbers from 00 to 99. I have an action and an object associated with each character so that I can use the PAO system, lumping three pairs, six digits, at each location of the memory palace.

The PAO system generates a unique image for every possible set of six digits from 000,000 to 999,000—that is, a million possible images. I couldn't possibly remember them all but every training session offers me unique stories to visualise on the fly.

Look at row one above, which starts with 2 and then 1 as the first pair. In my system these numbers, 21, stand for RJ, which are my late father's initials. The next number is 43, which is Alexander Bell. His action is to telephone someone. The third pair is 22, RR, Roger Rabbit. He eats carrots. So the first six digits give me the image of my father telephoning somebody on a carrot. I then imagine Dad performing this ridiculous act in the first location of one of the memory palaces I use for numbers. If I am using the palace which starts on a railway bridge, for example, Dad is sitting on the edge making his phone call using a carrot.

I move on to the next six digits in the next location, and so on. It takes seven locations to memorise the first row. The last location is only four digits, and so only requires a person and an action. In this case, it is 77 and 15, Tom Thumb and Johann Sebastian (Bach). The tiny boy is conducting an orchestra in the supermarket toilets.

In the Australian National competition, there were two Random Number events, one allowing five minutes for memorising, the other fifteen minutes. In 2017 I managed

68 digits in the five minutes but only 100 in the fifteen-minute event due to penalties. In 2018, I messed up in the five-minute event, but improved my fifteen-minute tally to 132.

In both competitions, a number of competitors scored zero even though they had memorised plenty of numbers. The challenge was whether to move as fast as possible and risk inaccuracy, or move slowly, remembering fewer numbers but reducing the risk of scoring zero. I chose to play it safe, knowing I am not good under pressure and was often careless in training. This strategy worked for me. I'm sure that had I gone at my maximum speed, I too would have scored zero.

In international events, athletes are given one hour to memorise in Random Numbers, which is a test of accuracy and endurance. In the Speed Numbers event they are only given five minutes, a test of speed.

The third event involves memorising numbers as they are spoken, one number per second. This is often considered the hardest event. I consider it impossible. I tried to place an image for each number in a memory location but was never fast enough to have fixed one number before the next two or three had already been called. I just remembered the first six numbers and blocked my ears until I could write them down.

The world record for the Random Numbers event at the time of writing is 3238 digits in one hour by American Alex Mullen. Mullen also holds the record for Speed Numbers, having memorised 568 digits in five minutes. For the horror event, Spoken Numbers, the record is an inconceivable 456 digits by fellow American Lance Tschirhart.

Sidetracking to memorising pi

Memory athletes love to memorise the digits of pi. It isn't a competition event, but they do it anyway.

Rick de Jong is a volunteer with the Dutch Alzheimer's Foundation. His father and three uncles had the disease. Wanting to raise awareness of Alzheimer's and demonstrate the memory power of a healthy brain, Rick decided to memorise the digits of pi: 3.14159265358979323846264338327950288419716939937510582097494459230781640 6 . . . and I'll stop there. (The 3 at the beginning doesn't count in the tally. They figure you already knew that one.)

In 2014, Rick started training for over an hour every day and more on the weekends. In the last few weeks before the recitation, he practised for a couple of hours a day. In his first trial, after about three months of training, he got to digit 4891 before making a mistake. That's it—one mistake and the trial was over. Rick tried again a few months later. He reeled off 5055 digits and forgot the next one. All over, once again.

In March 2015, he recited 22,612 digits to take the European record. It took him 5 hours and 37 minutes.

I asked Rick about the method he had used. He said that he started with something similar to the Dominic System, with a Person–Action–Object format, but moved on to his own method of using images interacting with each other at specific locations. He stored six digits at each location. After memorising 4800 digits, he adjusted his system to give him eight digits per location. His record required him to use over 3000 locations, recalling the image of a story at each one, which he would then convert to numbers.

Rick mentioned something very important for all memorisation: 'for deeper memorisation I don't just see these images but I give the new story a meaning, a kind of explanation. This rationalisation of a strange story helps my memory a great deal.'

Dominic O'Brien offered the same advice—try to make images rational, with a good reason for happening. It certainly helps to remember them, but it is slower to try to rationalise the really weird combinations that come up in competition. There is always a balance to be found between speed and accuracy. For memorising pi, it is only accuracy that matters.

The rules for records of pi on the Pi World Ranking website state that performances of 10,000 digits or more must be certified by an academic, scientific witness or an official such as a lawyer, notary or minister. They must be present to certify the performance.

Five hours of reeling off numbers is a long time and I suspect you have the same question as I do: what about going to the toilet?

After every 5000, I took a break for a drink and something to eat. And to go to the restroom. The judge accompanied me when I went to the toilet to make sure there was no cheating and I didn't write all the numbers on the bathroom wall.

I asked Rick why he would want to memorise the digits of pi. Alongside his personal ambition to get a record, he added:

I also wanted to learn more about long-term memory from personal experience. I think the experience in 'useless' memory training helps me when I memorise useful things. In regular memory sports it's all short-term memory and I'm personally very interested in long-term memory. Nevertheless, I also enjoy the competition. I learn a lot from my fellow memory athletes.

And if you think Rick's 22,612 digits is astounding, then take a moment to consider Rajveer Meena from India, who recited pi for about ten hours until he'd managed 70,000 digits, taking the world record in March 2015. He only kept it for seven months—fellow countryman Suresh Kumar Sharma then went 30 digits better. It took him seventeen hours and fourteen minutes. At the time of writing, Sharma's record stands.

Rick says he's thinking about going for the world record again. But there's also all the practical stuff he wants to remember, as well as training for memory competitions. And most importantly of all, there's his work with the Alzheimer's Foundation.

I much prefer to spend my time memorising practical knowledge too, but I had to have a go at pi. I've memorised the first 1000 digits and think that might be enough for me.

Memorising strings of 1 and 0

My favourite number event involves memorising strings of binary numbers—that is, 0 and 1 in rows of 30 digits. I have no idea why it is my favourite. I just like it. Again, the first error reduces your score by half and the second reduces the score for that row to zero. The start of the sheet for the competition will look something like this:

001110010100100000000110100011   Row 1
011110100011111110010100101110   Row 2
000100100000001000101110101001l   Row 3

I group the binary digits in threes and assign the letter I have used in the Dominic System. So 000 is 'O', 001 is 'J', 010 is 'R', 011 is 'B', 100 is 'A', 101 is 'S', 110 is 'G' and 111 is 'T'. Each pair of letters gives me the character from the Dominic System, and again I use Person–Action–Object.

The first row above gives me 001 (J) 110 (G) 010 (S) 100 (A) 100 (A) 000 (O) 000 (O) 110 (G) 100 (A) 011 (B). This becomes JG (Julia Gillard) doing SA's action (SantA, filling a sack) with AO (Annie Oakley's object, a gun). Australia's ex-prime minister stuffing a sack with guns, a highly memorable image you must admit, is placed in the first location of the memory palace. The rest of the row, OG AB, gives me Oscar the Grouch from Sesame Street doing the action of Alexander Bell, making a phone call. That image goes into the second location of the memory palace, and I move on to the next row.

My best score is 111 binary digits in five minutes in competition, going steadily and slowly to ensure no mistakes. At the time of writing the world record for the 30-minute event is an astonishing 6270 digits. The record holder, teenager Munkhshur Narmandakh, competes as part of the young Mongolian Memory Team, alongside her sister, Enkhshur Narmandakh, and Mongolian-Swedish dual citizen and world record holder Yanjaa Wintersoul. The three young women are having a major impact on international memory competition, putting women competitors at the same rank as the men while showing what a small country can do when it takes this growing sport to heart.

Fictional Dates

In the Fictional Dates event, there is a page of random dates between the years 1000 and 2099 and alongside each a fictional event. The first few might look like this:

1965   Phone speaker mode silenced on train
2010   Man with leaf blower assassinated
1034   Bomb dropped on noisy party at 4 a.m.
2025   Everyone puts rubbish in bins

(There's some wishful thinking here, but you get the idea.) It is similar to the Random Numbers event. I use the Dominic System and the familiar old Person–Action–Object format. The first two digits give me the person, the second two give me the action and the event gives me the object. There are more efficient methods given the limited number of centuries, but I am sticking with something I can manage easily.

By this stage of the 2017 competition, I was really struggling with concentration and only managed ten dates in five minutes. Thirteen in 2018 wasn't much better. The world record is held by Alex Mullen, who memorised 133 dates in the same time.

Names and Faces

The Names and Faces event consists of pages of head shots. Underneath each photograph is a given name and a family name. The organisers ensure that the names are culturally mixed so no one has an advantage. Consequently, most of the names will not be familiar. My best is eleven names in five minutes, losing lots of points for spelling. That has always been an issue for me.

The world record is currently 212 points in fifteen minutes, held by Yanjaa Wintersoul.

Random Images

The two memory championship associations differ in this event. The original competitions, organised by the World Memory Sports Council (WMSC), used rows of five abstract shapes filled with abstract textures for this event, and the newer International Association of Memory (IAM) prefers rows of five concrete images, such as a ball, a house, a young child, a star and so on.

The IAM organised the Australian event in 2017: in Random Images, I scored 40 images in five minutes. I improved to 75 in 2018, still far fewer than I could do in training. Both years, the images event was early on the second day and my ability to handle the pressure was fading fast. It only got worse.

The current world record is for the fifteen-minute event, in which Yanjaa Wintersoul scored 354 points.

Random Words

In Random Words, contestants are presented with a list of well-known words to memorise, in columns of twenty words with five columns to a page. The majority are concrete nouns with a few abstract nouns and a few verbs. One mistake costs half your points for that column. The second mistake means that column counts as zero.

In the 2017 five-minute event, I scored for only ten words, even though I had memorised many more than that. I'm still extremely cross at myself for the stupid mistakes I made. At the fourth location in my memory palace is a photograph of my father. The word was 'projector'. My father loved his slide projector and so I thought to myself—*Beauty! Easy!*—and moved on to the next location. When it came to recall, all I knew was that it was something my father really liked but could not recall the projector. I simply hadn't fixed the image properly. I made quite a few mistakes like that. Put it down to inexperience.

The world record is in the longer fifteen-minute event: 318 words by British competitor Katie Kermode.

The glamour event: Speed Cards

The most dramatic competitive memory event is memorising a single shuffled deck of cards as rapidly as possible. The maximum time is five minutes. If you cannot memorise the entire deck within that time, you score the number of cards you memorised by the five-minute mark. Although I memorised well over half the deck in 2017, I only scored eleven cards because you receive no further points after making a mistake.

I'm still unable to work out what I did. In training, I might mistake the Five of Clubs for the Five of Spades, say—an understandable error when working at speed. But in the competition, I apparently saw the Seven of Diamonds and memorised it as the Five of Clubs. I still cannot imagine how. But by the afternoon of the second day I was no longer handling the pressure—I just wanted to be anywhere but that silent room.

There are two trials of Speed Cards and your best score is recorded, but the second trial was even worse. I simply gave up after a few cards.

In 2018, I was again struggling with concentration and pressure. Although in practice I had broken the magical five minutes a few times, I knew that I was not capable of doing so in competition. I went slowly and scored 36 cards.

A week after the 2017 championships in Melbourne, the IAM World Memory Championships took place in Jakarta, concluding, as is the tradition, with the Speed Cards event. The competition was streamed live and I was astonished to watch Alex Mullen set a new world record, memorising an entire deck of cards in 15.61 seconds. Less than a week later, at the WMSC competitions in China, Chinese memory athlete Zou Lujian set a new record of 13.96 seconds. Mongolian Shijir-Erdene Bat-Enkh holds the record as I write, at 12.74 seconds. I cannot conceive how any of them can get their hands to move through 52 cards in that time, let alone commit their order to memory.

At the end of the three gruelling days in international competitions, the scores are tallied and manipulated by some complex mathematical algorithm that I will not even attempt to understand, let alone explain, and an overall champion is declared. At the time of writing, the IAM World Champion is Alex Mullen of the United States while the WMSC World Champion is Johannes Mallow of Germany.

Australian Memory Champion, Anastasia Woolmer
When I arrived to observe the 2016 WMSC Australian Memory Championships on the second day, at the invitation of memory athlete Daniel Kilov, organiser Jennifer Goddard

immediately took me aside to say a memory athlete no one had heard of before was really challenging Daniel for the lead. It was causing quite a stir. The competitor, Anastasia Woolmer, went on to be the first female to win the Australian title. She also set two new Australian records, memorising 360 binary digits in five minutes and 304 random digits in fifteen minutes.

Anastasia became a friend. She considers herself to have only an average natural memory. She left home and school aged fifteen to train at the Australian Ballet School. She has performed both classical ballet and contemporary dance in Europe and Australia. Anastasia left professional dance to have children and study economics, earning a prestigious Prime Minister's Australia Asia Endeavour Award. When she was nearly 40 she turned to memory training.

> While still fit physically, I began to feel my mental agility ageing in my late 30s. I set out on a training regime to regain a fit mind, focusing on memory techniques. Five months later I won the Australian Memory Championship, the first female to do so, and now feel my cognitive function is the best it has ever been.

Using publications and online forums as a guide, Anastasia set her own training schedule, drawing on the physicality of her background.

> I created my own system for ordering and memorising numbers, cards and binary. It has a strong emphasis on movement and directional changes associated with people

and objects. There is also no use of letters to assist in creating or memorising an image, such as in the Major System. Instead, it made more sense to me to be efficient. I devised my own code to take the visual cue of numbers, cards and binary sets and associate that with an image and its unique story and movement. My images have so much movement in them that I have taught this system in dance workshops with aspiring professional dancers, so they can create new choreography from number sequences.

Anastasia and I met up on the night before the 2017 IAM competitions. She was so pale from flu that she glowed faint yellow. Seeing her the next morning, I didn't expect her to survive the day. Although performing well below her best times, she still managed to be in the lead by the evening, and the next day set another Australian record in that horrible event, Spoken Numbers: she recalled 86 digits in order without error having only heard them read aloud, one number each second. I had stopped 80 digits earlier and watched her from the distance of a few desks. Anastasia won the national title convincingly for the second time.

Reflecting on her high marks in her undergraduate degree, Anastasia says that 'after learning memory techniques, I now know I could have achieved the same marks with half the study time'. She maintains that memory training, physical fitness and being creative are very much complementary.

Memory training has had a massive impact on my life. I feel liberated and confident as I face new challenges that I would have shied from in the past. I am able to take on new learning adventures with relative ease and I know I can learn anything I like. I feel I've been given the ability to learn two lives' worth of knowledge in one—it is fantastic.

As we age there is so much emphasis placed on maintaining our physical fitness and health in order to enjoy a long and happy life. I believe that mental fitness is just as important, probably more so. While memory training is virtually unknown, it is an amazing and very achievable path to mental fitness. With a little training almost anyone can perform apparently impossible memory feats.

The impact of training on concentration

My major concern when I first watched the memory championships in 2016 was that I would never be able to concentrate so intensely for the length of time required. I am so easily distracted. The first few times I attempted to memorise an entire deck of cards was painfully slow. I was constantly interrupted by my thoughts. *Are there better ways to lay out the cards? What was the noise outside? Why can't I remember the Jack of Clubs? Did I remember to get onions for dinner?*

I bought earmuffs. I had seen many memory athletes using them at the competition. The earmuffs reduced outside sounds, but it was the act of putting them on which mattered. The ritual became a signal to my brain to concentrate on memorising cards and numbers. I was able to catch my straying thoughts and force them back on track with more ease each day.

The idea of routine causing your brain to behave in a particular way was not new to me. Only a few weeks before I had talked to my GP about the struggles I had with sleeping. Poor sleep is the great enemy of memory and concentration—and health in general. She talked about how rituals trigger the brain into behavioural patterns. She set a sleep ritual. It was nothing radical; these ideas have been around for a long time. I had to turn off all electronic devices an hour before going to bed, drink warm milk and watch or read something gentle. Then, I was to go to bed and read something unrelated to work. Turning off the lights, I was to breathe deeply and just concentrate on the sounds around me. It worked a treat. After decades of battling poor sleep, I was gracefully slipping into slumber.

I started to note other factors that affected my concentration. Unexpectedly, a significant one was the temperature. If the room was 20 degrees Celsius, or even 19 or 18, my concentration was affected. I wore warm clothes but didn't heat the room above 17 degrees Celsius.

It seems obvious now, but at the time I hadn't thought about the impact of exercise. Dominic O'Brien suggested that I try doing some aerobic exercise before training. I walk every day and do other forms of exercise, but usually in one burst to get it over and done with. I am not an exercise enthusiast. But I got into the habit of putting on some lively music and dancing just before training. Again, the impact was significant, not only on my concentration but also on my enthusiasm for the cards and numbers.

Dominic was the first World Memory Champion in the inaugural event run by the World Memory Sports Council in 1991. He went on to win eight championships, a record that has never been equalled. After serving the sport in various administrative capacities globally, Dominic returned to the competitors' seats to take the World Senior Memory

Championship in China in 2017. In the same year, he won the UK National title, beating a swag of contenders many decades younger than his 60 years. He repeated the feat at the 2018 UK and Pan European events. Dominic's unique experience is revealing:

> I find competing in memory sports a way of combating self-doubt. Just when I fear my working memory is perhaps not as sharp as when I was younger, I am reassured by achieving even better results as I get older.
>
> You are never too old or too young to start training your memory.

I could not agree more.

APPENDIX A

Table of memory methods

	Method	Chapter	Example in this book	Suitable for
1	**Visual alphabet**	1	Public speaking	Temporary data, but also useful for anything that works like a list
2	**Bestiary**	1	Names	Useful for anything for which a word is significant but there is no requirement to remember an order
3	**Story, imagination**	1	Everything	The basis of all memory methods

	Method	Chapter	Example in this book	Suitable for
4	Memory palace (discrete)	2	Countries, periodic table, French verbs and genders, Chinese radicals	Almost anything as long as the knowledge can be structured. The most powerful and effective memory method known
5	Memory palace (continuous)	2	History Journey	Any data that does not have discrete items; for example, time, which is continuous
6	Bush songline	2	Calendar, seasons, natural environment	Linking memory palaces with the natural environment they are set in. For anything environmental or seasonal
7	Dance, song	2	Countries	Enhances all methods
8	Skyscape	4	None given as I have not found this a practical method for me	Works as a form of memory palace, but also works for timekeeping and navigation due to movement. Not easy to use in contemporary life and well-lit cities
9	Posts, poles and standing stones	4	Victorian mammals	Not very practical unless you have large natural stones or a post you can decorate

	Method	Chapter	Example in this book	Suitable for
10	Rapscallions	4	French genders	Adding conversation, emotions and narrative to abstract information. Characters add to the emotional impact of stories and therefore enhance all memory methods
11	Card decks	4	Historic people	Useful for sets of discrete data with a limited number of items
12	Dominic System	4	Dates	Anything with numbers
13	Lukasa-board-1	5	Field guide to birds	Just like a miniature memory palace and can be used for any data that can be structured
14	Stave or cane	5	Genealogies	Useful for any information with a linear structure
15	Memory boards	5	Spider families	There are many formats of memory boards from around the world that work like the *lukasa* above and can be used for almost any information that can be structured

	Method	Chapter	Example in this book	Suitable for
16	Landscape memory board	5	History	Enhancing landscape memory palace recall
17	Songboard	5	Anatomy	Singing knowledge makes it more memorable. The songs can be represented on a physical device
18	Body parts	5	Ancient mnemonists	Order not critical. Knowledge you want at hand at any time
19	Necklaces, bracelets and other beaded objects	5	Shakespeare's plays	Sequential data of limited length. The beauty of these objects is that you can wear them
20	Knotted cord *khipu*	5	Art and artists	These are a very adaptable aid including numeric data. They are best when physically present, e.g. when worn
21	Set of objects	5	Greek and Roman mythology	Wonderful for data with a complex structure that suits narrative rather than sequential memory systems
22	Hands	5	Astronomy science and history	Like a memory board, almost anything that can be structured, but has limited detail

	Method	Chapter	Example in this book	Suitable for
23	**Memory ball**	5	Ceremonial cycle	Any shape will work as a memory board, including carved spheres
24	**Lukasa-board–2**	6	Evolution of writing systems	Just like a miniature memory palace, but designed to suit the data, making it more effective
25	**Mandala**	6	Legal precedent, physics experiment	Like mind maps, mandalas are great when the information doesn't suit a linear layout but is related to a central theme or character
26	**Narrative scroll**	6	History of timekeeping	These really suit information that follows a narrative path, especially anything to do with history—recent or ancient, over short periods or long
27	**Drolleries and text decoration**	7	Medieval manuscript on musical instruments	Decoration of notes on any topic can make them more memorable, especially more abstract aspects
28	**Story illustrations with rapscallions**	8	Mathematical tables and equations, spelling and grammar	Abstract concepts are made memorable by giving them character and narrative

	Method	Chapter	Example in this book	Suitable for
29	**Winter count**	9	Personal history, twentieth-century history	Discrete events with limited number. Added to regularly
30	**Deck of cards**	10	Memory competitions	Use of memory palaces with added stress!
31	**Major system**	10	Random number lists	For memorising numbers for competition or general purposes

APPENDIX B

Bestiary

A suitable animal was not always possible, so some inventiveness was necessary.

Aa aardvark	As asp	Ch chipmunk
Ab Abyssinian cat	At atlas	Ci cicada
Ac acorn	Au auroch	Cl clown
Ad adder	Av avocet	Co cow
Ae aerialist	Ax axolotl	Cr crab (hermit)
Af Afghan hound	Ay aye-aye	Cu cupid
Ag Agaric fungi	Az Aztec	Cy cyborg
Ah Ah!—a sigh	Ba bat	Da dalmatian
Ai Airedale terrier	Be bee	De devil
Aj Ajax	Bi bison	Di diamond python
Ak Akita (dog breed)	Bl bloodhound	Do dolphin
Al alligator	Bo bower-bird	Dr dragonfly
Am amulet	Br brontosaurus	Du duck
An angel	Bu butterfly	Dw dwarf
Ap ape	By Byron	Dy dystopia
Aq aquatic leech	Ca camel	Ea earwig
Ar armadillo	Ce cenotaur	Eb ebb and flow

Ec echidna	He heron	Ma macaw
Ed edelweiss	Hi hippopotamus	Mc McScottish
Ee eel	Ho horse	Me Medusa
Ef effigy	Hu hummingbird	Mi millipede
Eg egret	Hy hyena	Mo monkey
Ei eider	Ia iambic	Mu muskrat
El elephant	Ib ibex	My mynah
Em emu	Ic Icarus	Na nautilus
En engraving	Id identical twin	Ne Neptune
Er ermine	Ig iguana	Ni nit
Es eskimo	Il illusionist	No noddy (bird)
Et ethereal	Im impala	Nu numbat
Eu eucalypt	In insect	Ny nymph
Ev Eve	Ir Iris	Oa oak
Ew ewe	Is island	Oc octopus
Ex Excalibur	Iv ivy	Od Odin
Ey eye	Ja jackass	Og ogre
Ez Ezekiel	Je jellyfish	Ok okapi
Fa falcon	Ji jigsaw	Ol olive
Fe ferret	Jo jonquil	On onion
Fi fish	Ju juggler	Op opossum
Fl flamingo	Ka kangaroo	Or orangutan
Fo fox	Ke kestrel	Os ostrich
Fr frogmouth	Kh khaki	Ot otter
Fu fuchsia	Ki kiwi	Ow owlet
Ga gaggle	Kn knight	Ox oxen
Ge genie	Ko kookaburra	Oy oyster
Gh gharial	Kr kraken	Pa panda
Gi giraffe	Ku kumquat	Pe penguin
Gl glider	Ky kylin (dragon)	Ph phoenix
Gn gnome	La lapwing	Pi pig
Go gorilla	Le leprechaun	Pl platypus
Gr griffin	Li limpets	Po polar bear
Gu gull	Ll llama	Pr praying mantis
Gw gwardar	Lo lobster	Pu puffin
Gy gypsy	Lu lute	Py python
Ha hare	Ly lynx	Qu quail

Ra raven
Re reindeer
Rh rhinoceros
Ri ringtail
Ro rooster
Ru runner bean
Ry rye
Sa satyr
Sc scorpion
Se seahorse
Sh shrew
Si Siamese cat
Sk skunk
Sl sloth
Sm smuggler
Sn snail
So sourpuss
Sp spaniel
Sq squirrel
St stingray
Su sugar glider

Sw swallow
Sy symphony
Ta tapir
Te termite
Th thornbill
Ti tiger
To toad
Tr troll
Tu turkey
Tw twitcher
Ty tyrannosaurus
Ul Ulysses
Um umbrella bird
Un univalve
Ur urchin
Us usher
Ut utopia
Va vampire bat
Ve Venetian
Vi viper
Vo volcano

Vu Vulcan
Wa walrus
We weevil
Wh whippet
Wi witch
Wo wolf
Wr wren
Wy wyvern
Xa xanthorrhoea
Ya yachtsman
Ye yeti
Yo Yorkshire terrier
Yu yucca plant
Yv scroll apology
Za Zapotec
Ze zebra
Zi zither
Zo zombie
Zu zucchini
Zy zygote

For all others use the visual alphabet:

A Arachne
B bird of paradise
C cat
D dragon
E eagle
F frog
G goat
H Hydra
I imp

J jester
K kitten
L lion
M marmoset
N Neanderthal
O owl
P panther
Q Quetzalcoatl
R rat

S skull
T toucan
U unicorn
V vulture
W wombat
X Xena, warrior woman
Y yak
Z Zeus

APPENDIX C

Prehistory Journey

Many of the dates given here have been updated or are the matter of much debate. The Prehistory Journey gives me an idea of the chronological order of events and reasonable dates. I update the dates by adding to the stories when I become aware of new research.

Time (general)	Geologic eons and eras	Time (specific)	Geologic periods and epochs	Examples of events occurring in the time period
START	HADEAN EON			The moon formed
4500 mya				
4000 mya	ARCHEAN EON			First life, photosynthesis
2500 mya	PROTEROZOIC EON			Eukaryotes—first multicellular animals emerge during this eon
540 mya	PHANEROZOIC EON Paleozoic era	540,000,000		First land plants
		530,000,000		Cambrian explosion—fast rate of evolution, 540–520 mya
235 mya	Mesozoic era	235,000,000	TRIASSIC PERIOD	Dinosaurs
		201,000,000	JURASSIC PERIOD	
1st CORNER 145 mya		145,000,000	CRETACEOUS PERIOD	
125 mya		125,000,000		Flowering plants appear
66 mya	Cenozoic era (geologic era till present)	66,000,000	PALEOGENE PERIOD Paleocene epoch	
		65,500,000		Major event—killed the dinosaurs

Time (general)	Geologic eons and eras	Time (specific)	Geologic periods and epochs	Examples of events occurring in the time period
		56,000,000	Eocene epoch	
		34,000,000	Oligocene epoch	
23 mya		23,000,000	NEOGENE PERIOD Miocene epoch	
		15,000,000		*Pierolapithecus*—common ancestor Hominidae, great apes
		6,000,000		*Orrorin tugenensis*—Millennium Man, Kenya
5.3 mya		5,300,000	Pliocene epoch	
		4,000,000		*Australopithecus* 2 spp.
		4,000,000		*Australopithecus afarensis*
		4,000,000		*Kenyanthropus platyops* (Lake Turkana, Kenya)
		3,300,000		*Archaeological time scale begins overlapping with the Geological time scale*
		3,200,000		Lucy—*Australopithecus afarensis*, Ethiopia, 1974

Time (general)	Geologic eons and eras	Time (specific)	Geologic periods and epochs	Examples of events occurring in the time period
		2,700,000		*Paranthropus* spp., robust australopithecines
2.58 mya		2,580,000	QUATERNARY PERIOD Pleistocene epoch	
	ARCHEOLOGICAL TIME SCALE			
		3,300,000	Start Lower Paleolithic	
		2,600,000		*H. habilis*
		2,000,000		*H. erectus*
		2,000,000		*H. georgicus* (Georgia, first outside Africa)
		2,000,000		*H. ergaster*
		2,000,000		*H. erectus*: Turkana Boy, Kenya
2nd		1,000,000		*H. erectus* in China
CORNER				
1 mya				
900 kya				
800 kya				

Time (general)	Geologic eons and eras	Time (specific)	Geologic periods and epochs	Examples of events occurring in the time period
700 kya		700,000		*H. pekinensis*, Peking man, near Beijing (or *Homo erectus pekinensis*)
600 kya		600,000		*H. heidelbergensis*
500 kya		500,000		*H. heidelbergensis*, Boxgrove, UK, West Sussex
		430,000		*Homo neanderthalensis* or *Homo sapiens neanderthalensis*
400 kya		400,000		*H. neanderthensis*
		315,000		*H. sapiens*
300 kya		300,000	Start Middle Paleolithic	
200 kya		200,000		
		150,000		Mitochondrial Eve
		125,000		Peak Eemian stage interglacial
		120,000		*H. sapiens* left Africa, mtDNA haplogroup N, C, A?
100 kya		100,000		
		90,000		Y chromosomal Adam

Time (general)	Geologic eons and eras	Time (specific)	Geologic periods and epochs	Examples of events occurring in the time period
		75,000		Toba volcano eruption
		70,000		Blombos Cave
		65,000		Kakadu
		60,000		Humans to Australia
50 kya		50,000		Lake Mungo
		50,000		Humans to Near East, Haplogroup B
		42,000		Paleolithic flutes in Europe
		41,000		Denisova hominins, Siberia
		40,000	Start Upper Paleolithic	
3rd CORNER 40 kya		40,000		Cro-Magnon colonisation of Europe
		35,000		Zar and other caves, Azerbaijan
		32,000		Aurignacian culture in Europe
30 kya		30,000		Chauvet cave
		30,000		Haplogroup X, I

Time (general)	Geologic eons and eras	Time (specific)	Geologic periods and epochs	Examples of events occurring in the time period
		30,000		Bow and arrow
		28,500		Humans to New Guinea
		28,000		Humans to Japan
		26,000		Start Last Glacial Maximum
20 kya		20,000		
		18,000		End Last Glacial Maximum
		18,000		Altamira
		17,000		Lascaux
15 kya		15,000		Humans to Americas
12 kya		11,700		End last Ice Age and end of Pleistocene
		11,000		Göbekli Tepe
		10,800		Start Younger Dryas
4th CORNER 10 kya		10,000		Burrup Peninsula = Murujuga
		10,000		Natufian culture
		10,000		Barley, wheat, Mesopotamia, now Iraq

Time (general)	Geologic eons and eras	Time (specific)	Geologic periods and epochs	Examples of events occurring in the time period
8000 BCE		9700	Start European Mesolithic	Start of Holocene epoch in geologic time scale
		8000 BCE		End Younger Dryas
		7500 BCE		Çatalhöyük, Turkey
		7150 BCE		Cheddar Man, Cheddar Gorge, United Kingdom
7000 BCE		7000 BCE		Jiahu, China
6000 BCE		6000 BCE		Iberian megalithic cairns and plaques
5500 BCE		5500 BCE		Copper, Pločnik, Serbia
		5500 BCE		Agriculture started in Ancient Egypt
		5500 BCE		Beginning of the Xinle culture in China
5000 BCE		5000 BCE		Start of Neolithic constructions in Orkney
		5000 BCE		Goseck circle, Germany

Time (general)	Geologic eons and eras	Time (specific)	Geologic periods and epochs	Examples of events occurring in the time period
		5000 BCE		Wheel
		5000 BCE		Protowriting
		4700 BCE		Menhirs and mounds, Brittany, France
4500 BCE		4500 BCE		Nabta Playa, Egypt, stone circles built
		4500 BCE		Maltese temples
		4000 BCE		Hal-Saflieni Hypogeum, Malta
4000 BCE		4000 BCE		Mesopotamian civilisations
		3800 BCE		First agriculture, Windmill Hill, Avebury
		3800 BCE		Post Track, Sweet Track—causewayed enclosures
		3700 BCE		Minoan culture starts
3500 BCE		3500 BCE		First mummies in Egypt
		3500 BCE		Watson Brake, Louisiana, United States

Time (general)	Geologic eons and eras	Time (specific)	Geologic periods and epochs	Examples of events occurring in the time period
		3300 BCE		Newgrange, Knowth, Dowth, Ireland
		3100 BCE		Standing Stones of Stenness; passage cairns, Orkney
3000 BCE		3000 BCE		Stonehenge started
		3000 BCE		Stone rows, Carnac, France
		3000 BCE		Avebury henge constructed, United Kingdom
		2700 BCE		Minoans; Knossos
		2600 BCE		Caral-Supe, Peru
		2600 BCE		First pyramid, Egypt
		2560 BCE		Great pyramid, Egypt
2500 BCE		2500 BCE		Ring of Brodgar, Orkney
		2500 BCE		Skara Brae, Orkney
		2100 BCE		Xia Dynasty, China (c. 2100–c. 1600 BCE)
2000 BCE		2000 BCE		Pre-classic Maya (to 250 CE)
		1800 BCE		Poverty Point, United States

Time (general)	Geologic eons and eras	Time (specific)	Geologic periods and epochs	Examples of events occurring in the time period
1500 BCE		1500 BCE		Shang Dynasty, China (c. 1700–1046 BCE)
		1500 BCE		Stonehenge abandoned
		1400 BCE		Olmecs emerge, Mexico
		1400 BCE		Lake Condah Aboriginal eel traps and stone huts, Victoria, Australia
		1323 BCE		Tutankhamun died, Egypt
		1200 BCE		Chavin de Huantar
		1200 BCE		Trojan War (very approximate date)
END 1000 BCE		1000 BCE		Start of History Journey

APPENDIX D

My chosen ancestors

Some of the dates in this appendix list are debated. I have quoted those I found most consistently when researching the characters. Spellings also vary. I did try hard to include more women but the selection criterion was those I could learn *from*, not just *about*. Regrettably, history did not often record the words of women in earlier times and I need those words to learn from. The History Walk has a better gender balance, with an abundance of women and men to learn about. I admit that my personal biases show in this list—physicists and mathematicians are overrepresented, while other fields of endeavour should be more prominent. But this is my list of chosen ancestors.

Chosen ancestor	Birth	Death
1 Homer	c. 800 BCE	unknown
2 Pythagoras	570	c. 495
3 Confucius	551	479
4 Herodotus	484	425
5 Socrates	470	399
6 Plato	c. 428	c. 348
7 Aristotle	384	322
8 Alexander the Great	356	323
9 Euclid	c. 300	unknown
10 Archimedes	287	212
11 Cicero	106	43
12 Julius Caesar	100	44
13 Cleopatra	69	30
14 Augustus	63	14 CE
15 Jesus	4 BCE	30/33 CE
16 Pliny the Elder	23 CE	79
17 Ptolemy	90	168
18 Constantine the Great	272	337
19 Augustine of Hippo	354	430
20 Attila the Hun	406	453
21 Muhammad	570	632
22 Charlemagne	742	814
23 Averroës	1126	1198
24 Genghis Khan	1162	1227
25 Fibonacci	c. 1170	1240
26 Thomas Aquinas	1225	1274
27 Dante Alighieri	1265	1321

Chosen ancestor	Birth	Death
28 William of Occam	c. 1287	1347
29 Petrarch	1304	1374
30 Geoffrey Chaucer	1343	1400
31 Johannes Gutenberg	1398	1468
32 Mehmed the Conqueror	1432	1481
33 Pachacuti Inca Yupanqui	1438	1471
34 Christopher Columbus	1450	1506
35 Leonardo da Vinci	1458	1519
36 John Major	1467	1550
37 Niccolo Machiavelli	1469	1527
38 Nicolaus Copernicus	1473	1543
39 Michelangelo	1475	1564
40 Sir Thomas More	1478	1535
41 Martin Luther	1483	1546
42 Henry VIII	1491	1547
43 Charles V, Holy Roman Emperor	1500	1558
44 John Calvin	1509	1564
45 Miguel de Cervantes	1547	1616
46 Francis Bacon	1561	1626
47 William Shakespeare	1564	1616
48 Galileo Galilei	1564	1642
49 Johannes Kepler	1571	1630
50 Thomas Hobbes	1588	1679
51 René Descartes	1596	1650
52 Oliver Cromwell	1599	1658
53 Blaise Pascal	1623	1662
54 Louis XIV, the Sun King	1638	1715

Chosen ancestor		Birth	Death
55	Isaac Newton	1643	1727
56	Gottfried Leibniz	1646	1716
57	Daniel Defoe	1660	1731
58	Voltaire	1694	1778
59	Benjamin Franklin	1706	1790
60	Carl Linnaeus	1707	1778
61	Leonhard Euler	1707	1783
62	Jean-Jacques Rousseau	1712	1778
63	Denis Diderot	1713	1784
64	Adam Smith	1723	1790
65	Immanuel Kant	1724	1804
66	James Cook	1728	1779
67	Catherine the Great	1729	1796
68	James Watt	1736	1819
69	Edward Jenner	1749	1823
70	Johann Wolfgang von Goethe	1749	1832
71	Wolfgang Amadeus Mozart	1756	1791
72	Napoleon Bonaparte	1769	1821
73	Ludwig von Beethoven	1770	1827
74	Jane Austen	1775	1817
75	Johann Carl Friedrich Gauss	1777	1855
76	Charles Babbage	1791	1871
77	Michael Faraday	1791	1867
78	Charles Lyell	1797	1875
79	John Stuart Mill	1806	1873
80	Abraham Lincoln	1809	1865
81	Charles Darwin	1809	1882

Chosen ancestor		Birth	Death
82	Otto von Bismarck	1815	1898
83	Karl Marx	1818	1883
84	Queen Victoria	1819	1901
85	Florence Nightingale	1820	1910
86	Fyodor Mikhailovich Dostoyevsky	1821	1881
87	Gregor Mendel	1822	1884
88	Louis Pasteur	1822	1895
89	Augustus Pitt Rivers	1827	1900
90	Marianne North	1830	1890
91	James Clerk Maxwell	1831	1879
92	Sitting Bull	1831	1890
93	Lewis Carroll	1832	1898
94	Pyotr Ilyich Tchaikovsky	1840	1893
95	Friedrich Nietzsche	1844	1900
96	Alexander Bell	1847	1922
97	Frances Hodgson Burnett	1849	1924
98	Oscar Wilde	1854	1900
99	Sigmund Freud	1856	1939
100	Nikola Tesla	1856	1943
101	J.J. Thompson	1856	1940
102	Emmeline Pankhurst	1858	1928
103	Max Planck	1858	1947
104	Sir Arthur Conan Doyle	1859	1930
105	Marie Curie	1867	1934
106	Gertrude Bell	1868	1926
107	Vladimir Lenin	1870	1924
108	Ernest Rutherford	1871	1937

Chosen ancestor		Birth	Death
109	Winston Churchill	1874	1965
110	Albert Einstein	1879	1955
111	Helen Keller	1880	1968
112	Pope John XXIII	1881	1963
113	Mustafa Kemal Atatürk	1881	1938
114	Benito Mussolini	1883	1945
115	John Maynard Keynes	1883	1946
116	Niels Bohr	1885	1962
117	Srinivasa Ramanujan	1887	1920
118	Erwin Schrödinger	1887	1961
119	Jawaharlal Nehru	1889	1964
120	Mao Zedong	1893	1976
121	Louis Leakey	1903	1972
122	Alan Turing	1912	1954
123	Paul Erdös	1913	1996
124	Nelson Mandela	1918	2013
125	Rosalind Franklin	1920	1958
126	Benoit Mandelbrot	1924	2010
127	Martin Luther King	1929	1968
128	Mikhail Gorbachev	1931	—
129	Kofi Annan	1938	2018
130	Linus Torvalds	1969	—

ACKNOWLEDGEMENTS

There are so many people who have helped with the preparation of this book that I am unable to thank them all. I would love to give details of all they contributed, but that would take far too long.

May I particularly thank Julia Adzuki, Felicity Albrecht, Tansel Ali, Stu Annels, Francis Blondin, Kirsten Boerema, Tessa Borgy, Calvin Bowman, Joseph Bromley, Catherine Carby, Rob Cas, Dino Cevolatti, Oliver Claycamp, Grace Coff, Kathryn Coff, Josh Cohen and community at the Art of Memory forum, Ian Colditz, Glenn Colville, David Cunningham, Harriet Cunningham, David Curzon, Evie Danger, Jacqueline Dark, Andy Fong, Leigh Franks, Ted Gioia, Sue Greed, Chris Groot, Duane Hamacher, Brendan Hanson, Joan Hooper, Ken Killeen, Daniel Kilov, Shufen Lin, Shulan Lin, Christopher Lincoln Bogg, Reuben Macdougall, Julie McHale, Meredith McKague, Di Manno, Susan Martin, Anthony Metivier, Lisa Minchin, Joan Newman, Julie Nihill, Dominic O'Brien, Nungarrayi, Mark de Raad, Jennifer Rodger, Xinxin Ruan, Lily Russell, Cameron Schmidt, Carolyn Tavener, Anastasia Woolmer and Tyson Yunkaporta.

I would also like to express my sincere thanks to the many readers of *The Memory Code* who wrote to me with such enthusiasm and shared their stories of implementing the memory methods mentioned there.

I would like to thank the wonderful staff and students at Malmsbury Primary School and Castlemaine Secondary College along with Indigenous colleagues from Nalderun and students from The Meeting Place. I would also like to thank the staff and students I engaged with at LaTrobe, Melbourne and Monash universities, and the University of Western Australia.

I could not have even attempted the art for the visual alphabet, bestiary, Rapscali tables, medieval manuscript and narrative scroll without the patience of my art teacher, Richard Baxter. He took on a total beginner with no natural talent and has produced an art enthusiast. I'd also like to thank my fellow art students Kerrie Peeler, Josh Smith and Calan Stanley for making classes so enjoyable. Thank you also to Suzanne McRae of Hip Hop Designs who created my special rapscallion, Rapscali.

I will always be hugely grateful to Tom Chippindall who spent many hours working through the practicalities of indigenous memory devices with me and helped create my personal versions of them. I am equally grateful to Paul Allen and Alice Steel whose enthusiasm for my research led to the foundation of The Orality Centre from which they run workshops and continue to research the implications of orality for education. Their insight has been invaluable, as has their friendship.

At Allen & Unwin, I am indebted to my publisher, Elizabeth Weiss, who has been an indispensable part of this book

ACKNOWLEDGEMENTS

from its inception. I greatly appreciate the skills of editorial manager Angela Handley, editor Aziza Kuypers and proof-reader Hilary Reynolds.

I am so fortunate to have had the support of family and friends through both the exciting and difficult times in this journey. In particular, I must thank some very special people in my life: Bec, Rudi, Abigail and Leah Heitbaum, and Win and Sue King-Smith.

And first and foremost (which is why I am putting him last), there would be no book without my partner in everything I do, Damian Kelly.

ABOUT THE AUTHOR

Dr Lynne Kelly is a science writer and Honorary Research Associate at La Trobe University. She has spent most of her life teaching physics, mathematics, information technology and general science at secondary school level. Lynne has written popular science books about her investigations into claims of the paranormal, crocodiles and spiders. During her PhD research on indigenous animal stories, she became fascinated with the way non-literate cultures manage to memorise a vast amount of practical information without writing.

On Salisbury Plain one day, looking at Stonehenge, she realised that these ideas offered a new way of looking at archaeological sites around the world. So began the exciting journey leading to *The Memory Code*, her sixteenth book. Lynne spent many years experimenting with memory methods from non-literate cultures alongside those from early literate societies and modern memory champions. It is her conviction that these techniques have a valuable role to play in contemporary life that motivated her to write *Memory Craft*.

NOTES

Chapter 1

1   The primary references on memory methods from the Middle Ages and Renaissance Europe are Frances Yates (1992, originally published in 1966), *The Art of Memory*, London: Pimlico, and Mary Carruthers (2008, originally published in 1990), *The Book of Memory*, Cambridge: Cambridge University Press.

2   My visual alphabet and bestiary are available as a combined booklet from my website <www.lynnekelly.com.au>.

Chapter 2

1   J. Bradley (2010), *Singing Saltwater Country: Journey to the songlines of Carpentaria*, Sydney: Allen & Unwin.

2   J. Foer (2011), *Moonwalking with Einstein: The art and science of remembering everything*, New York: Penguin Press.

3   Yates, *The Art of Memory*, p. xi.

4   Yates, *The Art of Memory*, p. 3.

5   A.R. Luria (1968), *The Mind of a Mnemonist: A little book about a vast memory*, New York: Basic Books.

Chapter 3

1   Jeremy Hance (2015), 'Amazon tribe creates 500-page traditional medicine encyclopedia', Mongabay, <https://news.mongabay.com/2015/06/amazon-tribe-creates-500-page-traditional-medicine-encyclopedia/>.

2   D. O'Brien (2014), *How to Develop a Brilliant Memory Week by Week: 50 proven ways to enhance your memory skills*, London: Watkins Publishing, pp. 63–4.

3   Andy Fong is a memory trainer and can be found through his website <http://andyfong.hk>.

Chapter 4

1   *Kungkarangkalpa: Seven Sisters Songline* website and the *Alive with the Dreaming! Songlines of the Western Desert* project (2018) <http://sevensisterssongline.com/resources/>.

Chapter 5

1   W.R. Bascom (1980), *Sixteen Cowries: Yoruba divination from Africa to the New World*, Bloomington: Indiana University Press.
2   E.M. McClelland (1982), *The Cult of Ifa Among the Yoruba*, London: Ethnographica.
3   Gary Urton (2017), *Inka History in Knots: Reading khipus as primary sources*, Austin, TX: University of Texas Press.

Chapter 6

1   James Shaheen (Spring, 2005), 'Tantric Art: Maps of Enlightenment', *Tricycle Magazine*, available at <https://tricycle.org/magazine/tantric-art-maps-enlightenment/>.
2   Zhang Hongxing (2013), *Masterpieces of Chinese Painting, 700–1900*, London: V&A Publishing, pp. 21–2.
3   Plato, quoted in Yates, *The Art of Memory*, p. 52.
4   Saint Augustine, quoted in Yates, *The Art of Memory*, p. 60.

Chapter 7

1   Yates, *The Art of Memory*, p. 93.
2   J. Burchill (1962), translation of Thomas Aquinas' *De memoria et reminiscentia*, point 348, <http://dhspriory.org/thomas/english/MemoriaReminiscentia.htm>, accessed 24 June 2018.

Chapter 8

1   K.H. Basso (1996), *Wisdom Sits in Places: Landscape and language among the Western Apache*, Albuquerque: University of New Mexico Press, p. 33.
2   The complete booklet, *Rapscali's Mathematical Tables*, is available from my website, <www.lynnekelly.com.au>.
3   *Mullen Memory* is at <https://mullenmemory.com>.

Chapter 9

1   Kasper Bormans, a PhD student with the team at the time, explains their work in a fascinating TEDx Leuven talk, available on the TEDx Leuven website at <http://tedxleuven.com/?q=2013/kasper-bormans>, accessed 25 June 2018.

INDEX

Numbers in *italic* indicate photographs or illustrations

Augustine of Hippo 53, 56, 165–6,
170, 181, *Pl. 23*
Australian Aboriginal *see* Aboriginal,
Australian
Australian prime ministers 125
Averroës 93, 178
Aztec 140

ball
carved 119–20
ceremonial cycle *118*, 119–20, 125
Neolithic *118*
Basso, Keith 210
Baxter, Richard 27
Bayeux Tapestry 158
bestiary 22, 23–38, 267–9, *Pls 8–10*
Bible 24, 47, 82–3, 170
verses 174
binary 237, 250–2
birch bark scroll 107
birds 21–2, 109–15, 119
call 114
dance 114
families 110–15
taxonomists 115
body
memory device 130–6
tally system 130–1
Boerema, Kirsten 200–1, 203
Bogg, Christopher Lincoln 64
Bradley, John 35
Bradwardine, Thomas 16–17, 22, 182
brain *see* neuroscience
Brazil 63
British kings and queens 55
Bromley, Joseph 197
Brooklyn Museum 143
Buddha 148

Camillo, Giulio 43–4, *43*
Canadian First Nations 146

cane *see* stave
canon tables 173, *Pl. 22*
Capella, Martianus 175
Carby, Catherine 202
card decks 4, 93–4
card tricks 205
Carpentaria 35
Carruthers, Mary 24
carved balls 119–20
Castlemaine Secondary College
200
ceques 36, 140
ceremonial cycle 119–20, 125–6
ceremony 35, 107
championship *see* competition
characters 6–7, 33, 88–105, 175–6;
see also rapscallion
French 72–5, *73*
chemistry
periodic table 130, 189, 192–5,
194
radioactive decay chains 129–30
Cherokee 153–4
China 55
Damaidi characters 155
dynasties 53, 55
handscrolls 83, 156, 157–8
Jiahu symbols 155
oracle bones 156
writing 84, 155–9
Chinese language 4, 62, 80–7
characters 84–5, 87
dictionary 80
memory palace 85–7
radicals 85–6
script 84
songs 119
Chippindall, Tom *118*, 119–20, 143
Christianity 23; *see also* Bible
churinga 107
Cicero, Tullius 33, 178–9, 217–18